EMPLOYEE ENVIRONMENTAL INNOVATION IN FIRMS

To my husband, Didier, and my son, Clark

Employee Environmental Innovation in Firms

Organizational and managerial factors

CATHERINE ANNE RAMUS
University of California, Santa Barbara
Donald Bren School of Environmental Science and
Management

Routledge
Taylor & Francis Group

LONDON AND NEW YORK

First published 2003 by Ashgate Publishing

Reissued 2018 by Routledge
2 Park Square, Milton Park, Abingdon, Oxon OX14 4RN
711 Third Avenue, New York, NY 10017, USA

Routledge is an imprint of the Taylor & Francis Group, an informa business

A Library of Congress record exists under LC control number: 2002103118

ISBN 13: 978-1-138-72714-4 (hbk)
ISBN 13: 978-1-315-19098-3 (ebk)

Contents

List of Tables

List of Figures

Preface

The purpose of this research was to identify which organizational and supervisory support factors can positively influence employees to try to promote environmental initiatives in businesses. Based on the empirical work of Amabile, Conti, Coon, Lazenby, and Herron (1996), which showed that employee creativity in the workplace is encouraged by organizational and supervisory signals, we have developed our conceptual model (Figure 2.1). Built on findings from an extensive study of the organizational behavior literature in the areas of innovation, organizational behavior, and the environmental management literature, our model had two major parts. It used employee knowledge of/belief in management commitment to environmental policies as the signal of organizational commitment, testing which of thirteen environmental policies influenced employee eco-initiatives. And, for the supervisory support signal, it used six characteristics found in organizational behavior literature (innovation, competence building, communication, information dissemination, use of rewards and recognition, and management of goals and responsibilities), testing which of six sets of BARS influenced employee eco-initiatives. We analyzed a data sample composed of 353 mid and low level employees from European companies from a range of industries in order to empirically prove or disprove our eight hypotheses.

The data were analyzed using multiple logit analyses, Chi-square test of differences, likelihood ratio tests, and descriptive statistics procedures. The results confirmed that employee knowledge of their company's written environmental policy and their belief that their company is committed to such a policy was not only statistically significant, but strongly increased the probability that the employee will have tried to promote an environmental initiative in the work place. And, the results showed that supervisory behaviors that support environmental innovation, environmental competence building, environmental communication, use of environmental rewards and recognition and management of environmental goals and responsibilities, indeed have a significant positive influence on employee willingness to promote an environmental initiative. Eleven of the thirteen environmental policies and one of the behavioral BARS (environmental information dissemination) were shown to have no significant influence on employee eco-initiatives.

Employees who felt strong signals of organizational and supervisory encouragement were more likely to have developed and implemented creative ideas that positively affected the natural environment. Specifically, employees who perceived the existence of an environmental policy in their organization, and learning organization-type support behaviors from their supervisors, were more likely to have tried to promote eco-innovations. But interestingly, our results also indicate that supervisors, even in our positively-biased sample of European companies with strong environmental commitment, used less supportive behaviors when managing environmental as compared to general business activities. Furthermore, our statistical results show that environmental support behaviors from supervisors are a better driver for employee eco-initiatives than more general support not aimed at environmental management, although even these environmental behaviors have a differentiated impact. And, we found that environmental policies may have indirect, or conditional effects, which influence employee perceptions of supervisory support for eco-innovations.

Acknowledgements

I began work on this research in 1996 while at the International Institute for Management Development (IMD). Ulrich Steger, with whom I worked, gave me the freedom to explore the question of how organizations either support or fail to support employee environmental innovation. He was the ideal hands-off manager, allowing me to delve into the topic and explore it in a very self-driven manner. His insights were always helpful to me and I thank him for his support.

Many of the companies involved in testing the survey questionnaires and from which I collected employee data were involved in IMD's environmental management project, MIBE. The environmental managers in MIBE helped me to make the research relevant to them, while at the same time respecting the need to test the hypotheses in an academically rigorous manner. I thank them for the commitment they gave to exploring this topic with me.

This research became the topic of my doctoral dissertation at the University of Lausanne's Ecole des Hautes Etudes Commerciales (HEC). Alexander Bergmann, who was my thesis chair at HEC, pushed me to think critically about my research. His thumbprint is on this document in that he encouraged me to work and rework my thinking until it became a robust and well-argued written output. I thank him for engaging in this mutually painful and very fruitful process.

Maia Wentland Forte sat on my dissertation committee, and was an invaluable support in the research process. She understood the import of the work and encouraged me in the writing process. Hers is a face that I see smiling at me as this book goes to press.

Others that I wish to acknowledge who have had an important role in reading and critiquing earlier drafts of the present work include: Gordon Adler, Thomas Bateman, John Ehrenfeld, Xavier Gilbert, Mark Starik, Bart Victor, and Monika Winn. I also wish to thank Hugues Pirotte who helped with data analysis and Derek Sweatt who helped with the formatting.

Throughout the process of researching employee environmental innovation (ecoinnovation) I received wonderful support from my husband, Didier Cossin. I am grateful for his constant love and encouragement. And finally, I wish to thank Clark Cossin, my son. He has given me an invaluable gift – the reminder that there are more important things in life than research!

List of Abbreviations

BARS	Behaviorally-Anchored Rating Scale
BCSD	Business Council for Sustainable Development
CCPA	Canadian Chemical Producers' Association
CEC	Corporate Environmental Commitment
CERES	Coalition of Environmentally Responsible Economies
CIA	Chemical Industry Association
DDT	Dichlorodiphenyltrichloroethane
EMAS	Environmental Management and Auditing Scheme
EMS	Environmental Management Standard / System
EUROPIA	European Petroleum Industry Association
ICM	Integrated Chain Management
IMD	International Institute for Management Development
ISO	International Standard Organization
IV	Independent Variable
GLS	Generalized Least Squares
LCA	Life Cycle Assessment / Analysis
NRTEE	National Roundtable on the Environment and the Economy
OLS	Ordinary Least Squares
SETAC	Society of Environmental Toxicology And Chemistry
UK	United Kingdom
UNCED	United Nations Conference on Environment and Development
US	United States
WBCSD	World Business Council for Sustainable Development

Chapter 1

Research Problem and Literature Review

1.1   INTRODUCTION TO RESEARCH PROBLEM

This book combines research on organizational behavior with that of environmental management to set the stage to an empirical investigation of the question of the extent to which organizational and supervisory support factors are effective in motivating employees to participate in environmental innovations that can increase the sustainability[1] of firms. Therefore, we first identify relevant organizational and supervisory factors which we then test to find which are effective at motivating employees' environmental initiatives and which do not have a statistically significant impact.

We assume that the transition toward environmental sustainability of businesses will be enhanced by the ability of organizations to implement creative environmental solutions from employees. This assumption has four component parts:

- The assertion that ecological sustainability of businesses is dependent on innovative solutions. It is supported by the eco-efficiency research demonstrating that sustainable business practices depend upon more efficient resource use in the manufacture and delivery of products and services (Fussler, 1996; Schaltegger, 1997; Steger, 1998a; WBCSD, 1996 a, b; Ytterhus, 1997).

- The assertion that innovation is the implementation of creative ideas from individuals and/or teams. It is underlined by the work environment literature (Amabile, Conti, Coon, Lazenby and Herron, 1996).

- The observation that any change needs to engage employees in change processes. This has been shown by the vast body of organization behavior literature (Kanter, 1983; Pasmore, 1994; Smith, 1996; Strebel, 1996, 1998).

- The assertion that, like in other types of change, the degree of organizational support for employee environmental actions determines the success of such efforts. This has been highlighted by the literature on environmental change (Barrett and Murphy, 1996; Wehrmeyer and Parker, 1995).

Building on the work environment research of Amabile (1988) and Amabile, Conti, Coon, Lazenby and Herron (1996), which shows that the level and frequency of creative behaviors is influenced by the social environment, we developed a survey instrument to test whether a set of organizational factors affect employee motivation to generate new ideas in the environmental management area. We focus our research on employee environmental innovations in environmentally proactive firms.

Proactive companies are defined as those who are not only "enlightened", that is who have strong corporate support which goes beyond compliance with environmental laws and regulations (Petulla, 1987), but who see environmental actions as strategic opportunities for competitive advantage. Monsanto is an example of such a company. It has stated that

> environmental goals and commitments are no longer seen as a cost of doing business. The environment is our business; it is as integral to the success of Monsanto as are new products, quality, service, a favorable return to our shareholders, and dedicated and talented employees (Smart, 1992; p. 236).

Proactive companies, like Monsanto, link their environmental business strategy to sustainable development "spotting opportunities for revenue growth in environmentally sustainable new products and technologies" (Magretta, 1997). Thus, when we refer to proactive companies in this book, we are looking at companies who say they see competitive advantage and development of sustainable products/technologies as being key components of their environmental strategies, leading them to actively support eco-innovation.

Influences on Employee Eco-Innovation

In the aftermath of the 1992 United Nations Conference on Environment and Development in Rio de Janeiro, Brazil, external influences on corporations to become sustainable, such as regulations, stakeholder pressures, and social desirability have led proactive companies to adopt policies and organizational mechanism that support employee eco-innovations. Davis (1991), Fussler (1996), and Hart (1997) indicate that eco-innovation is a means for corporations to become more environmentally and economically sustainable in their activities. Beard and Hartmann (1997) indicate that individual creative actions are the key to successful eco-innovation in firms.

Hostager, Neil, Decker, and Lorentz (1998) and Keogh and Polonsky (1998) indicated that organizational support is necessary to encourage employee environmental innovations. Hostager et al developed a model, which asserts that employee eco-innovations are influenced by both organizational and individual factors. The firm provides a signal of desirability and organizational incentives for employees to eco-innovate, but the individual brings with him/her the capacity (skills and capability) and the intrinsic motivation (values) to do so.

Our research looks at the factors that are controlled by the organization, but does not address the individual's existing capacity (skills and capability) and intrinsic motivation (values), both of which are factors that can be affected by organizational support, but which might also exist regardless of organizational actions. Factors of environmental skills and resources provided by the company can also influence employee willingness to eco-innovate and have been included in our study in the behaviorally anchored rating scales of the survey.

Eco-Initiatives as a Proxy for Eco-Innovation

An environmental initiative is defined in our study as any action taken by an employee that s/he thought would improve the environmental performance of the company. Employee eco-initiatives can improve company performance by either decreasing the environmental impacts of the company (such as recycling, pollution prevention, etc.); solving an environmental problem in the company (such as reducing the need for hazardous waste disposal, eliminating chemicals that are harmful to worker health/natural environment, etc.); or developing a more eco-efficient product/service (such as increasing resource efficiency by producing a less energy intensive product, replacing a product with a service, etc.).

Eco-initiatives are a proxy for eco-innovation in companies. Whereas it might be difficult for an employee to judge whether or not his/her initiative was innovative, any employee can know with certainty whether s/he has ever tried to promote an eco-initiative. In our study we are interested in self-identified actions that improve company environmental performance, therefore, we asked the respondents if they had ever promoted an environmental initiative, a simply worded question that is unlikely to create philosophical or definitional problems for employees.[2]

Eco-initiatives can take place at any level in the company. In our definition, we are not concerned with the size of the impact of the eco-initiative, but rather that the employee tried to promote an environmental change, no matter how small or large, which s/he believed would improve the company's environmental performance. It is important to note that our survey was designed for the average company employee, many of whom work in blue-collar jobs on the shop floor. And, the survey was conducted in four different languages with employees from twelve different countries. Therefore, the definitions and language used in the questionnaire needed to be understandable to people of different levels of education, sophistication, and from different cultural backgrounds.

Note that one of the limitations of our study is the use of self-reported eco-innovations as our dependent variable.

Eco-Innovations Differ From Other Innovations

Eco-innovations are different from other innovations in that they require a higher degree of managerial attention and commitment, but they are similar to other innovations in that they require the same types of organizational and supervisory support. The reason that they require a greater degree of support is that environmental innovations are not considered central to the profitability of a company, whereas other innovations are seen as contributing to the raison d'être of the business. Environmental management is a relatively new business process where measurement of impacts is not easily related to the bottom line since costs of environmental impacts are difficult to measure (Steger, 1993).

And, our research provides evidence that the types of support necessary to encourage individual innovation in other areas are the same as are necessary to encourage individual *eco*-innovations. The general organization literature indicates that organizational and supervisory encouragement support employee creativity and innovation, and environmental management researchers like Milliman and Clair (1995), Shrivastava (1996), Stern (1992) and Storen (1997), use case study evidence to assert that these same factors can encourage employee environmental participation and eco-innovation. The unique contribution of our study is that it empirically tests which of the identified factors have a statistically significant impact on employee eco-innovation.

Our Study of Organizational and Supervisory Encouragement for Employee Eco-Innovation

Based on our review of the organizational behavior and environmental management literature, we have chosen to focus our empirical work on the issue of which organizational and supervisory factors support employee eco-innovations. Organizational support was defined in our study as encouragement of employee actions signaled by environmental policies and top management communication of commitment to those policies. We differentiate this type of support from management or supervisory support, which we defined as daily behaviors perceived by employees to encourage eco-innovation. We look at which environmental policies employees perceive to signal organizational support, and which supervisory support behaviors employees perceive to signal management support, and whether some of these factors have a positive and significant impact on employee willingness to promote an environmental initiative in the workplace.

Conceptual Model

We devised a conceptual model, which forms the basis for this survey instrument, postulated that the relationship between the relative strength of the signal from supervisors and the organization regarding protection of the natural environment, has a direct impact on the willingness of employees to promote environmental initiatives. Furthermore, the model uses the organization behavior literature and interviews at companies as the basis for defining those supervisory behaviors that support employee creativity.

Approach

In this book we present the results from an employee survey. The first part of the survey asks employees for their perception of company environmental policies. The second part, comprised of behaviorally-anchored rating scales (BARS), asks employees to assess the behaviors they observe from their supervisors. Traditionally, BARS have been used by managers to assess employee work performance and behaviors. In our research, the BARS instrument has proven an effective approach to gather data from employees about management behaviors they have observed that either support or fail to support environmental creativity and innovations. This is a new use for the BARS tool. We then fed the results of this survey into logit analyses in order to interpret statistically our results.

Research Methodology and Sample Selection

We chose our research methodology based on interviews and meetings with environmental managers from major corporations in Europe. Employees from six major European companies were the focus of our research. We studied large corporations since they are more likely to have environmental policies than small and medium-sized enterprises (Fischer and Schot, 1993; Hutchinson and Hutchinson, 1995). Our database included 353 low and mid level employees from twelve different countries.

Our sample was carefully constructed to include employees working in twelve different countries, in six different industries and in companies that employ more than 1500 employees. (Companies in the sample employed between 1500 to 41,000 workers, not including contract employees, who in some cases added significantly to the work force of the company.) The company who employed 1500 permanent employees also employs approximately 2300 contract employees each year. And it maintained the same environmental policies and performance standards for those contract employees as for its permanent employees. Therefore we believe that the companies in the sample are representative of multinational corporations with 3000 plus employees, with head-quarters in Europe and proactive environmental policies.

Each of the companies in the sample experienced different levels of external pressures for environmental protection based upon their headquarters location and area of operations. (For instance the oil company faced a high level of stakeholder scrutiny and regulatory pressures, whereas the entertainment company faced a relatively low level of stakeholder scrutiny and external pressure on its environmental performance.) The companies in the survey also experienced different market pressures based upon the products/services they provided. This diversity of national cultures and industries allows us to argue that the chosen sample of employees is representative of a larger population of employees working for international companies.

Unique Aspects of Our Research

First, while the general organizational literature has tested the factors of organizational encouragement and supervisory behaviors on innovation, and the environmental management literature asserts that these same factors will influence employee eco-innovation, there does not exist an empirical study that we know of which proves that these factors indeed apply to the environmental innovation case. In fact, we find that while authors like Milliman and Clair (1995), Shrivastava (1996), Stern (1992) and Storen

(1997) all assert, using case study evidence, that factors like environmental policies, training, performance appraisal, rewards/incentives, information systems, targets, and supervisory inspiration will encourage employees to try eco-initiatives, these assertions have not been tested for their statistical validity. While we agree with the assumption that some of these factors may be important, it is not possible to see from the preceding environmental management literature which of these factors were most important, and which were least important, or not at all relevant. Our research allowed us to address this question by interpreting the statistically significant results of an empirical survey.

Second, we come up with some interesting results that differ from the assumptions made in the environmental management literature. For instance, when we tested the environmental policy factors, we found that employee knowledge that the company supported specific sustainability policies did not have a significant influence on employee eco-initiatives in eleven out of thirteen cases. And, when we tested the six BARS factors of innovation, competence building, communication, information dissemination, rewards and recognition, and management of goals and responsibilities, we found that supervisory support for eco-initiatives using information dissemination did not have a significant impact. Thus, our study suggests that not all the factors that the literature says influence employee eco-initiatives are actually important. Furthermore, by testing the interaction between our two sets of independent variables (policies and behaviors), we find a complex relationship which had not previously been discussed in the literature.

Third, our work makes a connection that has not previously been made between the general organizational behavior literature and the environmental management literature. Specifically, we test whether the same factors identified as affecting employee creativity, willingness to innovate, and willingness to participate in the general literature are also those that affect employee environmental creativity, innovation and participation.

Fourth, we use the BARS tool in a new way. To our knowledge, behaviorally-anchored rating scales (BARS) have not previously been used for employee evaluation of supervisory behaviors. Rather, there is a long tradition, documented in the academic literature by authors such as Farr, Enscore, Dubin, Cleveland, and Kozlowski (1980), Landy and Farr (1983), Schwab and Heneman (1975), and Smith and Kendall (1963), of using BARS for managerial assessment of employee job performance. Certainly, we have found no use of BARS in the environmental management literature, nor have we found documentation of the use of BARS for employee assessment of managerial work behaviors in the general behavioral science literature. Thus, to our knowledge, it is accurate for us to assert that this is a unique use of the BARS measurement tool.

Structure of the Book

This book consists of five chapters. In Chapter 1 we discuss the literature relevant to our work. This chapter has three parts. In the first part we look at pressures on firms that have led them to support employee eco-innovation in order to move closer to achieving sustainable operations. In part two of Chapter 1 we look at the organizational behavior literature which indicates those factors which support employee motivation and willingness to innovate. Then, in part three we discuss what makes environmental management different from general management, and show that the same factors which support other innovation are those that are expected to support employee eco-innovation. Note that while we consider this literature review as useful to provide the context and background for our research, information on the empirical study begins with the conceptual model development in Chapter 2. (The reader only interested in our actual research development and results can skip directly to Chapter 2.) Thus, Chapters 2 through 5 provide the entirety of our study.

In Chapter 2 we present the conceptual framework for our research. In part one we present the organizational and supervisory support components of the framework. In part two we present the literature that supports our choice of environmental policies and define those policies. In part three we define the supervisory behavior categories (i.e. behaviorally-anchored rating scales (BARS)).

Chapter 3 describes the methodology used in our survey questionnaire development, discusses our criteria for companies and employees included in the survey, and data analysis tools used.

Chapter 4 presents an analysis of our results including the use of logit tests, Chi-square test of difference, likelihood ratio tests, and other methods of data analysis. In this chapter we present the results for the eight hypotheses we tested as well as an analysis of the interaction between the two sets of independent variables (policies and BARS).

Chapter 5 summarizes our general findings, identifies possible implications for research and practice and suggests future areas of research.

1.2 INTRODUCTION TO LITERATURE REVIEW

In this section we present the pertinent environmental management and organizational behavior literature. We have organized this literature into three parts. We determine in part one that employee eco-innovation can help environmentally proactive firms to become more sustainable (a goal that has evolved due to external pressures to protect the natural environment). We then follow the logic that the factors that firms use to support other types of innovation may be those that affect employee eco-innovation. Therefore, in part two we are interested in finding out how firms generally encourage innovation, and in part three we look at the literature that describes how firms support *eco*-innovation. Out of this three part literature review we develop a model of factors affecting employee willingness to innovate which forms the basis for our conceptual model.

Part one answers the question "why have firms engaged in eco-innovation?". Here we identify the external context of companies from an historical perspective, indicating that businesses have come under pressure to protect the natural environment. This has led to a search in environmentally-proactive firms for sustainable business practices. The literature in this section supports our assertion that employee eco-innovation is important for firms who are striving to do business in a more environmentally-sustainable manner.[3]

Part two of the literature review looks at organizational behavior literature to examine the issue of how firms motivate employees to innovate. We ask the question "what have firms done in order to support employee motivation and willingness to innovate?". From this literature we define an original model of factors that affect employee willingness to innovate, which is at the foundation of our empirical work. Also from the literature we can see that two types of support, organizational and supervisory, are both of importance to the development of a work environment where employees are willing to innovate.

In the third part, we move from the question of what organizational behavior research indicates as important for general work place support of innovation to the issue of organizational and supervisory support for *eco*-innovation. We ask the questions "what differentiates environmental management from general management?" and "are the same organizational and supervisory support factors applicable to the case of employee environmental innovation as in the case of other innovation?".[4] We find that the environmental management literature indicates that employee environmental innovation requires the same type of organizational and management commitment and support as other innovation requires, confirming that the factors we have identified from the organizational behavior literature are applicable to the environmental case.

Part One: The Impetus for Eco-Innovation in Firms

From a historical perspective, corporate environmental policy and responsibility have evolved to their current state largely as a result of external pressures. This study is not focused on proving which external pressures, be they regulatory, societal or economic, have caused proactive environmental attitudes and behaviors on the part of companies. We are interested in identifying the external context which has, over time, led some firms to support a policy of sustainability, leading to organizational encouragement of eco-innovation.[5]

Here we explore the historical evolution of corporate reactions to external pressures to manage their environmental impacts. The pressures we describe created the impetus for firms to protect the natural environment, leading to the establishment of environmental policies. The most advanced of those policies include concerns of "sustainable development", and adopt notions such as "eco-efficiency". The literature indicates that environmental innovations are one means by which companies can achieve these policy goals, with minimal economic sacrifices.

Part one of this chapter is divided into four main sections.

- First, we briefly introduce the birth of the environmental movement, and describe the manner in which society became aware of industry as part of the environmental problem.

- Second, we describe the types of pressures that businesses responded to when developing their environmental programs, and industry organization responses to such public and regulatory pressures.

- Third, we examine historical trends in firm responses. In particular we look at developments in company environmental policies that resulted from the increased external pressures.

- Finally, we discuss the recent corporate focus on sustainability. We describe the role of eco-innovation in achieving a policy of sustainability and how some proactive companies are applying these principles.

Birth of the Environmental Movement and Awareness of Business Responsibility

Since the 1960s the environmental movement in the developed world has increased public awareness of environmental degradation, and businesses role in this deterioration of the natural environment. Shaping this movement, Rachel Carson's *Silent Spring* (1962) (translated into a number of European languages) did much to raise the awareness of the general populous of the disconnect between nature and industrial chemical pesticide products like DDT. Growth of this movement was slow at first. But by 1972, the Club of Rome (a group of academics) developed a model highlighting the unsustainability of current world growth, industrial production and use of resources in a book entitled *Limits to Growth — a Report of the Club of Rome* (Meadows, Meadows, Randers and Behrens, 1972).[6]

   A public debate, including government, business leaders, environmental groups, academics and individuals began to seek appropriate actions for businesses to take to prevent environmental harms. Economist E. F. Schumacher with his book *Small is Beautiful* (1973) began to try to find solutions to the problem of damage to the natural environment resulting from the activities of companies. Schumacher, recognizing problems caused by the existing economic model, prescribed an alternative economic model, which would internalize the costs to society from industrial activities. Books, like Schumacher's, illuminated the form of environmentalism called "naturalism", defined by as the belief in ecological holism and interdependence of nature and human relationships to nature (Shrivastava and Hart, 1994).

   Eventually a dichotomy of opinion regarding businesses role in protection of the natural environment emerged. Some authors, like Carson, distrusted companies' ability and motivation to participate in building a sustainable relationship with the planet, its living organisms and its resources. This opinion was also emphasized in books like Andrew McLaughlin's *Regarding Nature: Industrialism and Deep Ecology* (1993), which focused on the incompatibility of capitalism and nature. McLaughlin concluded that a non-anthropocentric approach is the only possible salvation for the planet. Another example is Theo Colburn, Dianne Dumanoski and John Peterson Myers' recent book, *Our Stolen Future* (1996), that argues that synthetic chemicals are responsible for the wide-spread and persistent existence of endocrine disrupters which are now poisoning human and animal health. Books of this type advocate "radical" environmentalism rejecting incremental reforms (Shrivastava and Hart,

1994). Authors like these call into question whether business can be trusted with protecting human health and the natural environment.

Other authors represented a more hopeful line of argument, believing that not only business is capable of creating planet-saving solutions, but that its leadership is paramount to future success. John Davis's *Green Business* (1991) built the case, based on the theory presented in Schumacher's *Small is Beautiful*, that companies can indeed develop in a sustainable manner. Davis (1991) argued that it is both an economic and societal imperative that businesses balance the long term needs of society with current resource use. Paul Hawken in *The Ecology of Commerce* (1993) argued that a redesign of the current industrial systems, coupled with a change in human behaviors and demands, can make a world where "commerce and environmental restoration are synonymous" (p. xvi). These authors believe that well-intentioned, environmentally-caring businesses and individuals can create a sustainable solution to the planets' resource scarcity and pollution problems.

There is no agreement in the literature as to whether or not mankind can create a clean, healthy planet and a non-abusive relationship with nature, but there is agreement amongst the above authors that business, as it is today, will need to transform itself if there is to be movement toward preserving and enhancing the natural environment. The "fringe" environmental movement of the 1960s became a middle-class movement in the 1990s through increased public awareness of the issues. Indeed, many opinion polls in the United States, Europe and the former Soviet Union show environmental protection as one of the most important issue on people's minds (Ehrenfeld, 1978; Landy, Roberts, and Thomas, 1990; Shrivastava and Hart, 1994). Thus, public opinion over the past thirty years has been increasing pressure on companies to take responsible actions to protect human health and the environment. (We will discuss the impact of this movement on businesses below.)[7]

Various Pressures on Corporations, and Opportunities From Protecting the Environment

Pressure on corporations to develop environmental protection programs came from a number of different sources, including customers, regulators, shareholders, investors, and environmental groups. And, at the same time as these pressures increased, some firms saw opportunities from protecting the natural environment. Fischer and Schot (1993) discuss four types of pressure/opportunities leading to corresponding changes in industry behaviors. The types of pressure fall into the categories of regulations, credibility, market and financial. Netherwood (1996) defines the pressures (and opportunities) on industry as legislative, green marketing-related opportunities, public pressure, ethical concerns and commitments to Agenda 21 (the international policy document produced at the 1992 United Nations Conference on Environment and Development). Others define the pressures as related to different pressure groups or stakeholders, including investors, shareholders, employees, environmental organizations, community groups, etc. (Belz and Strannegard, 1997; Winter and Steger, 1998). For what follows, we will use the four categories of Fischer and Schot's as they are largely inclusive of the others highlighted above. Note that while we describe these as "pressures", many companies have seen increased pressures as incentives to be proactive with regards to environmental protection.

Regulatory Pressures Regulatory pressures on firms resulted from increased public awareness of pollution leading to government legislation aimed at addressing the public's concerns and protecting public health and the natural environment (Rappaport and Dillon, 1991). Out of this increased public concern, we find that national and international governmental bodies (like the United Nations) began turning their attention to environmental pollution. In 1972 governments from countries world-wide explicitly recognized the need to address industrial impacts on the natural environment at the United Nations Conference on the Human Environment in Stockholm (*The Human Environment*, 1972). Governments in the developed world responded with environmental regulatory policies (Rappaport and Dillon, 1991; Shrivastava and Hart, 1994; World Commission on Environment and Development, 1987). For instance, throughout the developed world regulatory environmental agencies were created in the 1970s responding to the building concerns about and evidence of environmental harms (Shrivastava, 1996; Shrivastava and Hart, 1994). By 1977, more than sixty nations had established national environmental ministries, agencies, departments or secretariats (Gladwin, 1977). Growth of environmentalism, in both its radical and more moderate forms, manifest in new regulations, created tremendous pressure on business and industry to improve environmental performance[8] (Shrivastava and Hart, 1994; World Commission on Environment and Development, 1987). This performance mandate was not only evident in environmental regulations but also in the success of political "green" parties in countries like Norway, Sweden, Germany, and France, all of which made exacting demands on businesses to stop polluting (Porter, 1980).

By 1984 when the World Commission on Environment and Development was formed, there was agreement within most governments as to the international nature of environmental harms, which had previously been seen as a primarily local phenomenon. Evidence of the global impacts of environmental pollution included: acid rain, o-zone depletion, desertification, global warming or climate change, over-fishing of species, etc. The book resulting from this multi-national body's work, *Our Common Future* (World Commission on Environment and Development, 1987), defined Sustainable Development. In 1992, the United Nations Conference on Environment and Development held in Rio de Janeiro, Brazil resulted in international agreements encouraging governments and industry throughout the world to address environmental concerns as they continued to develop economically (Shrivastava, 1996). The conference directly linked business activities of corporations to world environmental degradation, and emphasized businesses role in solving ecological issues (Post and Altman, 1992).

Indeed the majority of European business managers responding to a five country survey agreed that national and international regulators were the most important pressure group for their business (Belz and Strannegard, 1997). Examples of the regulatory trend include the Large Combustion Plant Directive in the European Union leading to acid rain (sulfur emission) reduction legislation; the German Packaging regulation and other recycling and product take-back legislation in Europe.[9]

For our research it is interesting to note that the institutionalization of environmental regulations over time increased firms' motivation to integrate environmental management into their business processes. Eventually, a number of firms began to go beyond compliance with these regulations and to look for opportunities arising from proactively managing environmental issues in their businesses.

Credibility Pressures   Credibility pressures resulted from society not trusting companies' desire or ability to protect the public from their pollution (Fischer and Schot, 1993). We include in this category both events and groups that focused public attention on company environmental performance.

Examples of disasters leading to loss of trust include:

• Minamata, Japan where mercury contamination from the Chisso Chlori-alkali plant poisoned thousands between 1953 and 1960 in the local community who subsisted on fish from the polluted sea.

- Hoffmann-LaRoche's ICMESA chemical plant accidentally releasing at toxic cloud of dioxin over Seveso, Italy in 1976.

- Occidental Chemical's Love Canal, New York disaster, where hazardous waste near residential areas caused cancer, eventually leading to mass evacuation of homes in 1978.

- Union Carbide's Bhopal, India chemical leak that killed hundreds and harmed thousands in 1984.

- Sandoz's Schweizerhalle chemical plant catching fire, resulting in chemical pollution of the nearby Rhine River.

- Chernobyl, Ukraine nuclear disaster, resulting in radioactive contamination of people and land and world-wide loss of trust in nuclear fission as a source of energy in 1989.

- Exxon's Valdez oil tanker disaster in 1989 where a huge leak spilled tons of crude oil into the pristine ecosystem of Alaska's Prince William's Sound.

These widely-publicized, highly-visible ecological disasters lead to a loss of credibility for companies, who the public increasingly saw as unable to contain the externalities of their businesses.[10]

At the same time, environmental groups became active in destroying company environmental credibility through publicizing failures of companies to protect the environment, and through pushing governments to promulgate stricter environmental regulations and/or to enforce existing statutes.[11] Greenpeace (established in 1971), in particular, has successfully trained the international public eye on environmental issues through actions that have attracted media.[12] For examples Greenpeace has participated in nuclear plant protests, chasing of fishing boats who use drift nets, protecting of whales and baby seals, boarding of oil platforms scheduled for deep sea disposal (Shell Brent Spar). Others like World Wildlife Fund and Friends of the Earth have taken a softer approach, often protecting species and rain forests through appealing to public sensibilities (e.g. panda bears).

Also active in this trend of putting pressures on company credibility have been European-based environmental organizations. In 1970, German environmental activists pressed the German government to ensure that Chancellor Willy Brandt's election platform of environmental protection (in 1969) be taken literally (Bluhdorn, Krause, Scharf, 1995). And in the United Kingdom, the rise of environmental concerns in the late 1960s coincided with the expansion by both Labour and Conservative Governments of quasi-autonomous non-governmental agencies for the protection of the environment (Doherty and Rawcliffe, 1995).

While difficult to measure the precise effects, these environmental disasters and increasing power of activist groups have led to a tangible difference in business behaviors (Hawken, 1993). One certain effect is that companies now want to be better prepared to manage public relations battles, avoiding them whenever possible (Winter and Steger, 1998).

Alternatively, we can see a trend toward proactive actions from a small group of companies to develop an image connected to their environmental credibility. Acting to create positive environmental credibility through policies of sustainable development, firms like Patagonia, Body Shop, and Ben and Jerry's, have advertised their company environmental policies to the public when marketing their goods (Mirvis, 1996).[13] So, while some companies have tried to preserve their credibility in a reactive manner (avoiding disasters and reacting swiftly to address problems when they arise), others have seen environmental credibility as a means to gain business advantage and to position themselves in the marketplace.

Market Pressures Following Fischer and Schot's typology, we include below a discussion of both pressures and opportunities arising from the market. Market pressures arose from consumers who demanded environmentally-sound products and/or from demand for cleaner technologies. Some market pressure has been manifest in boycotts of specific products, like the World Wildlife Foundation's consumer boycott of tuna fished with nets that killed dolphins in 1991, and the Greenpeace's 1997 campaign against genetically engineered soy bean products. But while boycotts are effective at placing specific pressure on a particular product/company/industry, the larger trend in environmental marketing in Europe is still toward producers of eco-products turning toward "eco-niches" or green customer groups.[14] Stakeholder influence is increasing, thus opening "ecological mass markets" (Belz et al. 1997, Belz and Strannegard, 1997; also see Winter and Steger, 1998, for a thorough discussion of different stakeholder groups that influence companies. For instance, consumers, regulators, board members, employees, community organizations, etc.).

In a 1991 study of 592 German companies, 56.2 percent of the companies noted a growing market segment of environmentally-oriented consumers (Steger, 1993). Companies like Electrolux, Interface, Neste,

Patagonia, Philips, and Xerox, to name a few, began to redesign products to try to capture the market power of green consumers, who were often willing to pay more for greener products and services. (For example, a recent study in the United States showed that twenty percent of consumers were willing to pay up to ten percent more for energy from renewable sources (Xenergy, 1997).) But most of the time consumers expect environmental performance without being willing to give up design or to pay more to get it (Ramus, Steger and Winter, 1996). The trend is for consumer goods producers to pass this market pressure on to their suppliers. In a 1997 survey of businesses in Belgium, Norway, Sweden and Switzerland, 56 to 61 percent of companies surveyed had environmental procurement policies which place pressure on their suppliers to defend the environmental credentials of their products (Belz and Strannegard, 1997).

A proactive movement also arose. In the clean technology markets there has been a wide-spread increase in demand for environmentally-sensitive products and production. For example, there is growing commercial demand for solar power with the global market for solar energy increasing by 18 percent from 1995 to 1997, and projections that it will experience similarly large growth rate over the next five years (Park, 1997). Japan's Natural Resources and Energy Agency estimates that solar energy use will increase in Japan by 1000 times to 4.6 gigawatts by the year 2010 (Park, 1997). Another example of new markets that are creating incentives for clean technologies is that of zero-emissions vehicles. United States, Japanese and German automobile manufacturers are all focusing research and development into the creation of eco-vehicles using electric, fuel cell or other innovative technologies. In the United States, the market for environmental products and services is estimated to surpass the size of established industries like plastics and pharmaceuticals (Park, 1997). There is growing market demand for products and services with good environmental credentials giving an impetus to companies to fulfill these demands.

Financial Pressures Financial pressures come from banks, investors, and insurance companies who have become concerned about limiting liabilities from investments in companies with poor environmental performance. Many lending institutions in the North America and Europe began to require companies to perform environmental assessments on properties, as well as to fill out detailed environmental questionnaires to enable the financial community to assess future liabilities and predict future company financial performance based on environmental criteria (Ramus, Steger, and Winter, 1996). For example, since the advent of the Superfund regulation[15] in the 1970s it has become standard bank policy in the United States to have environmental impact assessments before lending on properties with possible hazardous waste liabilities. And, Union Bank of Switzerland and Deutsche Bank are two examples of banks who have recently announced that they are making company environmental performance a criterion for future lending and stock selection (Ramus, Steger, and Winter, 1996).

Insurance companies have also placed pressures on companies to manage their environmental impacts. Some companies began withholding coverage for environmental liabilities in the 1980s as a result of environmental disasters caused by companies (Hoffman, 1997). Insurance companies have become increasingly concerned about financial losses from climate change-related natural disasters and have reacted by putting financial pressures on companies (Park, 1997). Members of this industry were some of the most vocal proponents for an international agreement on climate change at the Global Climate Change Conference in Kyoto, 1997. Insurers have also begun looking at company environmental, health and safety performance when deciding on rates for premiums, reminding companies that there is a real cost from poor performance.

The availability of capital places another pressure on companies to improve environmental performance. The financial markets in the United States and Europe have begun to set up special environmental/social responsibility funds for individuals, pension funds, etc. who want to invest in environmental responsible companies and technologies. For example the Sustainable Asset Management fund (Switzerland), Environmental Management Fund (Norway), NPI Global Care Fund (UK) were all developed from 1995 in order to invest in companies with good environmental performance. Also in the mid-1990s Triodos Bank in the Netherlands started a savings account called Earth Saver Account which provides funds solely to environmental projects. In the United States, the Global Environment Fund (capitalized at 50 million US Dollars) invests in clean energy, waste water treatment and related sectors in emerging markets, and the Ventana Group's North American Environment Fund (capitalized at 50 million US Dollars) invests in promising firms specializing in air pollution control, hazardous waste management, and recycling (Park, 1997). Capital from these funds is distributed based on environmental performance and environmental product/service delivery, giving an incentive for companies to address environmental considerations in their daily business decision-making. This increased scrutiny from banks, investors and insurers has been one impetus for companies to focus on improving environmental performance.[16]

Industry Organization Responses to Pressures   By the 1990s industry organizations added an important impetus to individual company environmental protection activities. These organizations coordinated the public admission of corporate responsibility for environmental degradation, and by doing so increased pressure on businesses to develop policies to address environmental concerns. Industry organizations responded to the external pressures in stages. But when the above mentioned regulatory and societal awareness had reached a high point before the 1992 Rio Conference, the International Chamber of Commerce became proactive in their approach to environmental issues. In 1991, at the World Industry Conference on Environmental Management, industry presented a "Business Charter for Sustainable Development" which was signed by 600 firms world-wide. This represented recognition by many global enterprises that industry must maintain responsible economic development practices. And for the 1992 United Nations Conference on Environment and Development, Business Council for Sustainable Development (BCSD), an organization of 50 multi-national corporations, produced a book entitled *Changing Course*. This book gave a global business perspective on development and environment, at the conference, and it was used to represent industries' agenda for addressing environmental degradation at the conference (Schmidheiny, 1992). It proposed corporate responsibility and stated that proactive behavior on the part of corporations was the solution to the sustainable development issue.

Industry organizations began to take an important role in facilitating changes amongst their members (Park, 1997). Proactive businesses created the World Business Council for Sustainable Development in 1995 (a merger of the International Chamber of Commerce's World Industry Council on the Environment and the Business Council for Sustainable Development). This organization, a coalition of 120 international companies, was "united by a shared commitment to the environment and to the principles of economic growth and sustainable development" (WBCSD, 1996c). The companies in WBCSD work together proposing and implementing "solutions" to address the issue of human impact on the Earth's resources and nature's health.[17] The chemical industry associations in Canada, the United States and many European countries initiated the Responsible Care program to improve environmental, health, and safety of their business activities (Park, 1997). Keidanren (Japan's leading business group) and the Conference of Indian Industry adopted an environmental code of ethics for members (Park, 1997). Other organizations, like The Natural Step, founded by Karl-Henrik Robert in Sweden, established national organizations world-wide to educate businesses on the systems approach to sustainable actions. (Companies, like Electrolux, Interface, and Northumbrian Water, to name a few, use The Natural Step program to educate workers and managers about the environmental limitations that exist in the eco-system.) Through organizations of this type, industry has taken a serious and responsible role in shaping the solutions necessary to move businesses toward sustainable development.[18]

Firm Responses to Pressures

The above literature provides the general external context that led industry to take greater responsibility for its environmental impacts. This context of external pressure and industry group cooperation has led individual firms to develop policies of environmental protection. In this section we will discuss the trends in how individual companies have responded to environmental issues by progressively accepting their responsibilities for pollution resulting from their business activities. Below is a historical view of the evolution toward corporate responsibility resulting in environmental policies.

Following the birth of ecology in the 1960s, individual firms took actions to protect themselves from bad publicity surrounding their activities, and to comply with environmental laws. Eventually, some firms began to take a proactive approach to protecting human health and the environment. As stated above, policy changes in companies have been linked to regulatory, credibility, market and financial pressures. As these pressures have increased over time, so have corporate policies been put in place (Fischer and Schot, 1993; Hoffman, 1997; Shrivastava and Hart, 1994). Fischer and Schot in their introduction to the book *Environmental Strategies for Industry* (1993) give a historical perspective about firms' reactions to external pressures during the period of 1970-92. The authors tied external pressure to environmental management changes within companies describing two separate periods of activity, from 1970-85 and 1985-1992.

Regulatory Environmentalism   The period from 1970-85 has been described as a period of corporate policies that resisted adaptation to the pressures (Fischer and Schot, 1993, Gladwin and Welles, 1976). Hoffman (1997) calls this period (from 1970 to 1982) "regulatory environmentalism". These authors agree that during this period most environmental improvements in firms were done on an ad hoc basis, largely as a means of complying with government regulations. Larger firms (5000+ employees) were more likely to have formal environmental policy statements than small or medium-sized firms, who seldom had any formal policy during this period. This lack of policies is reflected in a 1974 multinational survey that found that less than 40 percent of 516 responding companies, including small and medium sized companies, had a formal statement of environmental policy (Gladwin and Welles, 1976). Even in large multinational corporations, environmental policies were developed and applied on an ad hoc basis corresponding primarily to local concerns about pollution. Thus, Dow Chemical's Global Pollution Control Guidelines tailored environmental policies to local situations (Dow Annual Report to Shareholders, 1972). And one European petroleum company's conservation advisor stated "our environmental policy and practice is adjusted to the different laws, attitudes, interpretations, esthetic values, and environmental conditions found in each nation" (Gladwin and Welles, 1976; p. 165).

The conclusion of the above cited research was confirmed by work done in this area by Petulla (1987), who conducted a four year study of corporate environmental policies and practices beginning in 1982 with firms in the United States.[19] As a result of this research Petulla divided corporate environmental policy development into three categories. 29 percent of the firms (mostly small-sized enterprises) had "crisis-oriented" environmental management. These firms had no environmental policy strategy for compliance with environmental laws and regulations. 58 percent of the companies had "cost-oriented" environmental management. These firms tended to be larger (1000+ employees) and made efforts to nominally comply with environmental regulations, often delaying the necessary changes as long as possible. Only 9 percent of the companies had "enlightened" environmental management, meaning that they had strong corporate support which went beyond compliance, included good environmental management practices, and often included environmental goals in their long-range plans. Top management support and the "quality and effectiveness of work in federal, state, and local agencies" were seen as key drivers toward more proactive policies (Petulla, 1987; p. 183). The early 1980s saw the majority of firms maintaining a reactive, locally-based policy approach, with some early movers beginning to see that environmental protection costs might be minimized if quality and cost control mechanisms were established.

Environmentalism as Social Responsibility This period was followed by a change of approach to environmental management between 1985-92, during which companies began "embracing environmental issues without innovation" (Fischer and Schot, 1993; p. 8). Hoffman (1997) calls a similar period, bounded by the years of 1982-1988, as "environmentalism as social responsibility". The authors agree that during this period the majority of firms were more rigorous in their incorporation of environmental considerations into company policies. For example, most firms had a formal written policy statement by the end of the 1980s (Flaherty and Rappaport, 1991; p.10). Sixty percent of companies in an Arthur D. Little survey stated they intended to go beyond compliance (Arthur D. Little, 1988). But, a study by Hunt and Auster (1990), concluded that most corporations were still in a reactive mode and that few corporations were at the proactive stage of policy development and implementation. And a 1991 study of German 592 companies showed that environmental protection was "only occasionally listed as an independent corporate objective" (Steger, 1993).

Company policy statements during this period often included requirements of their environmental, health, and safety programs, including emission/performance targets, and staff and facility responsibilities. The policy statements were increasingly used as internal and external communication tools with the internal purpose to put the environmental issues higher on the managers' and employees' priority list (Schot, De Laat, Van der Meijden, and Bosma, 1991). The focus of company environmental departments during this period was primarily on monitoring compliance and interfacing with external agencies and the public. There is general agreement in the academic literature that the strategic environmental management task was a minor part of company environmental management activities during this period (Post and Altman, 1992; Rappaport, Taylor, Flaherty and Pomeroy, 1991).

Leading up to the 1992 United Nations Conference on Environment and Development (UNCED), some efforts were being made to offer incentives for environmental protection in some company practices (Schmidheiny, 1992).[20] For instance, environmental performance was assessed using environmental audits, and environmental performance criteria were sometimes incorporated into line manager evaluations. Firms also began to integrate environmental impact assessments into planning processes for new facilities, and decision making (Fischer and Schot, 1993). The start of corporate environmental management systems was seen in companies during this time. Firms began evaluating environmental performance of businesses (Flaherty and Rappaport, 1991; p. 9). New product review procedures were beginning to be put in place in companies, but this process was used to minimize impacts, not to decide for or against product development in most cases (Schot, 1992). Most of these integration efforts were aimed at reducing pollution or increasing recycling rather than development of new alternative products (Schot, De Laat, Van der Meijden, and Bosma, 1991). Firms, during this period, began to incorporate responsibility for environmental protection into policy and operations, but firms were not yet grappling with issues of sustainable development. Environmental protection had not yet become a strategic issue for the vast majority of firms during this period.

Progression from Single-Loop to Double-Loop Learning   The above historical analyses have demonstrated that outside pressures have been internalized into corporate activities and philosophies in different ways. Some have seen environmental management as an addition to existing management processes and some have seen it as a means to achieving competitive advantage. There is a pattern by which most firms started by "adjusting" or modifying practices on an as needed basis to comply with regulations (beginning in the 1970s), a practice that results in incremental (single-loop) learning. Over time some, mostly larger, firms began a process of "adaptation" and "anticipation" resulting in a questioning of old assumptions (double-loop learning) which led to environmental protection becoming a part of core and basis business objectives (Post and Altman, 1996).

What is interesting for our research from the above literature is the fact that some firms decided to go beyond a reactive approach to external pressures, adopting policies of sustainability.[21] Indeed, a "rare set" of large companies have undergone a serious re-examination and re-evaluation of environmental goals with an intention of institutionalizing these goals in all parts of the company. Bringer and Benforado (1993) described this process as "environmental performance goals are clear, measurement is under way, and innovative opportunity-development is being systematically encouraged through a host of organizational incentives". While we know of no convincing examples of companies who have successfully evolved to a sustainable development strategy in all business areas, or along their entire value-chain, there are a few notable example of firms who are trying to do so (Ben and Jerry's, Body Shop, Interface, Monsanto, and Patagonia). Below we will look at how policies of sustainability[22] have led firms to look for tools to achieve these policy and business goals. The focus of many of these tools is the development of new environmental technologies, cleaner, safer, longer lasting products, and/or finding of innovative environmental solutions. Thus, support for eco-innovation is presented as a key means of achieving sustainable development in firms.

Eco-Innovation as a Means to Achieving Sustainability

The above analysis showed how regulatory, consumer, shareholder/investor and environmental group pressures had differing effects on corporations and their policies. A few firms after the 1992 Rio Conference stopped reacting to these pressures and began seeing environmental activities as business opportunities, using them as a source of revenue growth and competitive advantage. They began differentiating their businesses by producing sustainable products with sustainable technologies using environmental innovation as a strategy rather than simply complying with environmental requirements.

In this section we look at how companies are using tools to internalize concepts of sustainability and eco-efficiency into their activities. We do this in order to make the connection between proactive companies' desires to become more sustainable and their use of eco-innovations to achieve this end.

Many academics have written on the topic of corporate sustainability (Carpenter, 1993; Gladwin, 1992; Kinlaw, 1993; Milbrath, 1989; Nijkamp and Soeteman, 1988; Pezzey, 1989; Robert, 1989; Starik and Rands, 1995; Stead and Stead, 1992). These authors have tried to give insights into how companies can operationalize the concept of ecological sustainability in business practices. But, even with this large body of work, there has been a lot of confusion surrounding the issue of how companies implement sustainable management practices (Shrivastava, 1996; Shrivastava and Hart, 1994). Gladwin, Kennelly, and Krause (1995) provide a synopsis of the academic literature noting that sustainable development has been equated with so many different descriptions and prescriptions that the term is at a "rather high level of abstraction" (p. 878). While companies have been encouraged to become sustainable, seldom until recently have there been blue prints of what sustainable behavior would look like (Shrivastava, 1996).

Since sustainability is a topic that is surrounded by confusion for academics, we have chosen to focus on the practical aspect of this concept as a driver for change in corporations. It is our contention that beginning after the Rio conference in 1992 proactive corporations tried to operationalize sustainable development by inserting certain aspects of the concept into their internal environmental policies. This contention is supported by the literature and is discussed below.

Davis (1991) argues that innovations in the environmental area will be the key to attaining sustainable development. He states that "the future of sustainable development is one of intense innovation - inventing a new and different age. Innovation will be in the lead, with administration playing its supporting role" (p.27). And Kemp and Soete (1992) claim that some of the present technological trajectories have reached their environmental limits and need to be replaced by environmental-friendlier trajectories. Their research points to a need to develop and adopt new cleaner technologies, but they have identified technical, economic and institutional barriers to this change (Kemp and Soete, 1992). Furthermore, other researchers believe that in order to reverse the deterioration of global systems of climate control and sustain conditions that support human life the adoption of large scale technological change will be necessary (Heaton, Repetto, and Sobin, 1991; Speth, 1989).[23] Thus, we see agreement in the literature that eco-innovation (and the new technologies that result from eco-innovations) plays a central role in transforming companies into sustainable enterprises.[24] This is an important assumption underlying our research.

Environmental Technologies Shrivastava (1995b) defines environmental technologies quite broadly, including in his definition hardware technologies, the design of production equipment, new products and delivery mechanisms, as well as operating methods and procedures, research and development processes. In his definition, as long as the technology conserves energy and natural resources and minimizes the environmental impacts of human activity it is an environmental technology. Environmental technology requires a management orientation toward consideration of environmental impacts in business decision making (Shrivastava, 1995b).

Eco-efficiency An important concept, called "eco-efficiency", makes one linkage between management orientation toward a policy of sustainable development and product/process design which minimizes environmental impacts. Eco-efficiency was first introduced into the academic press by Schaltegger and Sturm (1990) but it was Schmidheiny's Business Council for Sustainable Development (BSCD) which made it a popular concept in 1992 with the publication of *Changing Courses* for the Rio UNCED conference. Schaltegger and Sturm (1990) defined the concept as the "ratio between value added and environmental impact added, i. e. the ratio between an economic and an ecological performance number". Thus "eco-efficiency" is the positive correlation between environmental and economic efficiency (Ytterhus, 1997). (Companies use this tool by applying a formula that helps them make decisions between the tradeoffs of one product, process or service versus an alternative.) Ayres (1995) described it as a concept close to Design for the Environment, because of its goal of designing products and processes so as to minimize environmental impacts, while at the same time increasing company environmental value creation. The core of eco-efficiency is "making more with less", meaning to reduce resource intensity (including waste, pollution, energy and other resource use) when producing outputs. The concept of eco-efficiency is a key corporate response to the global goal of sustainable development (BCSD, 1994; Ytterhus, 1997).

Eco-compass The Eco-compass, a tool available to integrate eco-efficiency considerations into business decisions, is described by Claude Fussler in his book *Driving Eco-Innovation* (1996). The eco-compass is a decision-making framework helping to illustrate the relative environmental trade-offs between alternatives. Specifically, this eco-compass aids in new products/services development, weighing six dimensions against one another. These six dimensions include potential risk to environment or health, resource conservation, energy intensity, materials intensity, revalorization (remanufacture, reuse, recycling) and extending the productive life of the product/service (Fussler 1996; p. 151). It can be used in a consultative process, which can include customers, suppliers, engineers, etc., to provide companies with information to help them become more sustainable on a case-by-case basis. Thus, Fussler asserts that the eco-compass can help companies to profitably produce sustainable goods by applying eco-efficiency principles and tools which promotes greater design efficiency in early phases of new product innovation.

The Natural Step Another tool that companies can use to move toward sustainable development is The Natural Step framework of four fundamental systems conditions that, in the organization's view, need to be met in order to have sustainable development. National organizations exist in Europe and North America to promote this "systems approach" thinking in businesses. The four fundamental systems conditions introduced by The Natural Step are as follows.

- The first condition is that substances from the Earth's crust can not systematically increase in the biosphere, meaning that fossil fuels, metals and other minerals can not be extracted at a faster rate than their redeposit.

- Second, substances produced by society can not systematically increase in the biosphere, meaning that substances need to be able to be broken down by nature. Persistent chemicals that are not found in nature need to be phased out.

- Third, the physical basis for productivity and diversity must not be systematically deteriorated, meaning that we can not diminish the productive capacity of the Earth or its biodiversity. Renewable resource use is indicated.

- Fourth, there must be a fair and efficient use of resources to meet human needs (Robert, 1989).

The outgrowth of this approach has been integrated chain management or life cycle management, an environmental management process which takes the systems limitations into consideration during company decision making (Cramer a, b, 1996). Necessary to achieving these systems conditions is the use of environmentally sensitive technologies.

Applications of Eco-innovation Tools to Help Companies Become More Sustainable During the last five years some companies have begun grappling with how to apply the above concepts in order to transform their businesses into more sustainable enterprises. Some companies have taken these tools and concepts and begun implementation of sustainable development policies. Monsanto, a chemical and biotechnology company, is using less polluting technologies and the company's vision of sustainable development to remake its business (Hart, 1997; Magretta, 1997). Patagonia produces its products with minimal impact on the environment, using recycled materials to produce its high-quality outdoors wear (Shrivastava, 1996). Body Shop produces and sells health care products in an environmentally-sensitive manner. Its stated policy of environmental and social equity includes hiring environmentally-minded employees, employing homeless people in London, and creating demand for rain forest products to improve the situation of indigenous people who are connected to the Earth[25] (Roddick, 1995; Shrivastava, 1996; Shrivastava and Hart, 1994). Electrolux and Monsanto have begun to integrate The Natural Steps' systems thinking (Sustainability Review, 1997). And eco-efficiency tools, like the eco-compass, are being employed in companies like Dow Chemical, Procter & Gamble, and Sony as one means for developing sustainable corporations and firms (Fussler, 1996).

While Cramer (1996b), Fussler (1996), Roberts (1989), and Shrivastava (1995b) describe tools that can be used to help businesses manage sustainability issues, these tools largely focus on product design considerations, and alone they do little to encourage environmental innovation (Beard and Hartmann, 1997). Beard and Hartmann discuss the need for creative thinking in order to achieve eco-innovations across organizational value-chains. They see individual creative actions as a key to successful eco-innovation in firms. Below we look at the literature that demonstrates which types of support are necessary for creative input from individuals.

In part one of the literature review we have traced the evolution of corporate actions and policies observing that external pressures have led firms to become more protective of the natural environment when doing business. Some firms have seen an advantage in becoming proactive in their environmental protection activities, leading them to look for creative solutions in order to achieve their sustainability goals. Eco-innovation is an important means for these companies to make progress toward these goals. It is this small group of relatively large companies who are of interest to us in our research as our empirical study looks at company support for environmental innovation.[26]

Since our study is interested in developing and testing a model of organizational and supervisory factors supporting employee participation in eco-innovation, we review the literature on how organizations, in general, support and motivate employees to innovate. We do this because we assume that the factors that affect other innovation may be related to those affecting eco-innovation.

Part Two: Literature on Organizational Behavior and Innovation

In part two we present the relevant organizational behavior literature to examine the issue of how firms motivate employees to innovate. In this literature, we are concerned with identifying the relevant organizational behavior dynamics where they apply to employee participation in innovation. We begin by using the literature to define the areas in which we focus our study. Then, we explore the questions of

- What types of organizations support employee innovation?

- How organizations that are successful at innovating encourage employees to engage in initiatives? and

- What is the role of supervisory personnel in encouraging employee participation in innovations?

From the answers to these questions we derive a model of factors that affect employee willingness to innovate. (Part three of this chapter, as explained in the introduction of this section, is concerned with the identifying the factors that could affect employee eco-innovation. General organizational barriers to innovation are described below in part two. Barriers to environmental change are discussed in part three.)

Creativity and Innovation Theory

In our research we are interested in identifying organizational conditions that lead employees to be motivated to create or innovate. Creativity and innovation are linked in the literature. Under conditions of social and technological change, people's creative efforts make an important contribution to organizational performance (Redmond, Mumford, and Teach, 1993, Amabile, 1988, Davis, 1986, Grossman, 1982, Taylor, 1963, Tushman and Anderson, 1986). Creativity is connected to innovation since all innovations begin with creative ideas (Amabile, Conti, Coon, Lazenby and Herron, 1996). Amabile and her associates (1996) define creativity as the production of novel and useful ideas and innovation as the successful implementation of those ideas within an organization. They assert that the social or organizational environment can influence individual creativity and innovation. They also note that individuals will be most creative when they are intrinsically motivated by the interest, enjoyment, satisfaction and challenge of the work, but that this intrinsic motivation can be undermined by extrinsic motivators that lead people to feel externally controlled in their work (Amabile, 1983, 1988, 1993). Thus, the organization in which an employee works influences the creativity of that individual and his or her willingness to innovate.

Amabile and her associates (1996) tested for relevant organizational factors affecting worker creativity and innovation in an empirical investigation. They concluded from their investigation that five organizational factors seem to play an important role in influencing creativity: challenge of the work, organizational encouragement to create or innovate in a given area, work group supports, supervisory encouragement, and absence of organizational impediments, such as bureaucracy and hierarchy. Of these, we have chosen organizational encouragement and supervisory encouragement to create or innovate as the areas that we are most interested in for our research since there is an extensive environmental literature that agrees that these impact employee eco-innovation. Organizational impediments are looked at in terms of how they manifest themselves in supervisory behaviors. Work challenge and work group supports, while important factors, fall outside the purview of our study.[27]

Organizational encouragement is defined by Amabile et al (1996) as

> an organizational culture that encourages creativity through the fair, constructive judgment of ideas, reward and recognition for creative work, mechanisms for developing new ideas, an active flow of ideas, and a shared vision of what the organization is trying to do (p. 1166).

In our research we are concerned with all of these aspects of organizational encouragement but we separate them. What we call the "organizational signal" is concerned with the shared vision for ecological sustainability, which is demonstrated by environmental policies. Other parts of Amabile's definition are included in the supervisory behaviors section of our conceptual framework as categories of "innovation", "communication", and "rewards and recognition". (Please refer to Chapter 2 for further information on our approach.)

Supervisory encouragement is defined by Amabile et al (1996) as "a supervisor who serves as a good work model, sets goals appropriately, supports workers, values individual contributions, and shows confidence in the workers" (p.1166). We have incorporated these aspects of supervisory encouragement into our research, but have expanded the definition beyond Amabile's to include supervisory support for competence building, use of non-hierarchical management behaviors, and behaviors that support open communication and information dissemination (in addition to behaviors that support innovation, communication, and rewards and recognition mentioned above). We expand the definition because both the general and environmental literature indicate that these additional aspects of supervisory support may influence employee eco-innovation.

What Types of Work Environments Successfully Support Employee Involvement and Innovation?

Since our research is concerned with testing for organizational and supervisory support of employee innovation, we are interested in identifying which types of work environments and organizations successfully support participation and innovation by employees. To begin with we look at the literature to determine barriers to employee innovation. Then we look at characteristics of organizations that are successful in their support of employee participation and innovation.

Barriers to Innovation As an example of organizational characteristics that can create barriers to employee participation we look at the work of H. Fayol and F. Taylor, who were the major authors and researchers in so called Scientific Management theories. In these "machine" model theories, employees are considered passive instruments, capable of performing work and accepting directions, but not of initiating actions. These theories looked at how to optimize worker contribution to efficient production, leading to work by "bureaucratic sociologists" like P. Blau, A. Gouldner, R. Merton, and P. Selznick. Scientific Management theory eventually led to control-oriented management in companies, defined by Lawler (1992) as bureaucratic and hierarchical, with management controlling employee work using vertical relationships.

In this control-oriented management approach, the relationship of the supervisor to the employee was authoritarian. In addition, management treated employees as "instruments" of production, and jobs were divided into small tasks that have little meaning to the employee. The workers, treated like part of the machinery, were seen as replaceable parts, with no influence over the work or vision of the organization's goals (Lawler, 1992).[28] The result of this type of management relationship to workers tended to be poor work satisfaction, apathy and low productivity because the work was repetitive and boring, and individuals felt their part in the organization was small and meaningless (Huczynski and Buchanan, 1991). We draw the conclusion from the Scientific Management theory literature that organizations that want to motivate employees to innovate and develop new initiatives must avoid restricting employee participation to repetitive jobs that alienate them from the organization. More importantly, the relationship between management and employees benefits from a less authoritarian structure.

Further work by Merton emphasized the demotivating aspects of authoritarian and bureaucratic approaches to managing employees. In Merton's model, the underlying relationship to the worker as an instrument used by an organization to achieve productivity, results in unanticipated consequences for the firm (Merton, 1936, 1940, 1945, 1947,1957). Merton describes the following causation

Top of organization demands control: ⇒ Creates emphasis on reliable behavior within organization: ⇒ Standard operating procedures implemented: ⇒ Checking on employees to ensure procedures are followed.

According to Merton, the result of this causation was that employees were evaluated independent of their individual achievement, rules replaced organizational goals, and the number of possible decisions got limited to a small set of options.   In addition, according to Merton, power relations between employee and supervisor negatively affected productivity as employees resented control and oversight.   Thus, top-down, control-oriented management tended to hamper employee creativity and search for innovative solutions (Cummings, 1965).

While Merton and Cummings' research was not oriented toward organizations who openly support innovation, we take a number of learning points from their work on bureaucracy. We see that the type of bureaucratic structures that existed in organizations studied by them limited incentives to innovate by evaluating individuals independent of their personal achievements, and decreased the search for alternative solutions, by limiting decisions to a small set of options. We also conclude that the relationship of the employee to management has important implications for innovation in organizations. Specifically that Merton and Cummings both note that employees resent or are hampered by control-oriented management behaviors.   Therefore, we assume, based on the results of Merton and Cummings' work that organizations which take a less hierarchical, less bureaucratic and less authoritarian approach, treating employees as rational cognitive beings with values, rather than passive instruments, will have more success motivating participation in innovations. We test this assumption in our empirical study.

Characteristics of High-Involvement Organizations We also looked at the literature that describes work environments that successfully support employee involvement. Employee involvement has been shown to have significant positive effects on organizational effectiveness (Conger and Kanungo, 1988; Lawler, 1992; Quinn and Spreitzer, 1997; Randolph, 1995). Thus, approaches to management have emerged that give employees greater participation in decision making in organizations, as participation can increase motivation which positively affects organizational effectiveness. This form of management, called participative or high-involvement management is fundamentally different from "control-oriented" management. In a control-oriented approach, work is simplified, standardized, and specialized, with supervision and pay incentives used to motivate individuals to perform their tasks well, leaving little or no discretion or influence to the employee (Lawler, 1992).   Participative management organizes work to be challenging, interesting, and motivating, with individuals at all levels in the organization given power to influence decisions (Lawler, 1992). Lawler (1986) noted that in high-involvement organizations employees require information sharing, knowledge of company goals and vision, power to influence[29] and rewards linked to performance in order to participate effectively. He also notes that training to improve performance as well as supervisor support and leadership are key to the transition into this type of organization. Thus, for our conceptual model, we note that information, goals (and policies that support vision), participation and influence in decision making, rewards, training and the role of supervisors are all key factors in organizations that encourage employee involvement.[30]

High involvement organizations use employee empowerment in order to operate. A great deal of recent scholarly work on empowerment exists commencing with an influential article by Conger and Kanungo (1988) who cite Bennis and Nanus (1985), Kanter (1979, 1983), and McClelland (1975) as suggesting that the practice of empowering (or giving responsibility to) subordinates is a principal component of managerial and organizational effectiveness. Conger and Kanungo (1988) link empowerment to expectancy theory[31] by asserting that empowerment improves employees' expectations that their efforts will result in a desired level of performance. The strength of a person's belief in self-efficacy (i.e. the personal belief that one is capable of successfully performing a given action), will effect the effort that person will expend toward an outcome and how long s/he will persist in the face of obstacles (Bandura, 1977). Training to acquire new skills can also make employees feel more capable and able to attain goals (Conger and Kanungo, 1988). Empowerment and training can help to make employees more motivated, effective in their jobs, and can encourage employee innovation. [32]

An employee's sense of self-efficacy can be positively or negatively affected by the organizational context. Specifically, management behaviors can act as a positive or negative role model which encourages or discourages employees.[33] Fault-finding, in particular, acts against self-efficacy feelings in employees (Conger and Kanungo, 1988). Empowering practices by supervisor, including openness of communication, goal setting, non-bureaucratic and non-hierarchical approaches, openness to employee participation in decision making, feedback and good information on performance and goals, openness to experimentation,[34] and exposure to learning opportunities are factors noted as influencing work perceptions of self-efficacy (effectiveness in ones work) and empowerment (Bandura, 1977, 1986; Conger and Kanungo, 1988; Kanter, 1983). Supervisors have a significant role in empowering employees (promoting their involvement and participation in job actions) noting that the factors listed above are important.[35]

Tying the above literature on general employee involvement to employee involvement in innovations, Peters (1991) asserts that management support and commitment are key factors in supporting innovation. Affirming the findings of the above literature on general employee involvement, Peters notes that non-hierarchical management structures and culture, open participation by employees in decision-making, management use of rewards and recognition to help employee motivation, effective goal setting and responsibility sharing with employees, and training on skills necessary to participate in the innovation process and on the issues on the innovation agenda. These dimensions that are highlighted in Peters' research are also supported by other authors like Harrison (1983), Kanter (1983), and Walton (1985).

Field research at Digital Equipment in Scotland by Buchanan and McCalman (1987, 1989) also supports the supposition that the above mentioned factors are responsible for an increase in product innovation, worker productivity and reduced time to market. Specifically their research finds that individual and team responsibility, supportive, non-directive management style, training to help develop both technical and decision-making/problem-solving skills, computer-assisted information systems to increase flow of information, early communication of goals and vision for desired changes, rewards including promotion potential and revised payment/salary policy, and an evaluation process which included review/assessment of behaviors as well as achievements, are necessary support factors for innovation.

The literature and the above mentioned field studies indicate that organizational context determines, to a large degree, whether employees will participate in achieving organizational goals and in innovation in a given area. (Our empirical study confirms this conclusion.) Control-oriented, hierarchical management approaches are less successful at encouraging employees to participate than high involvement, empowerment approaches. In addition, the literature indicates that both organizational support and supervisory support are important areas from which employees derive their sense of encouragement.

Supervisor's Role in Influencing Creativity and Innovation

As we have highlighted above, while new ideas ultimately come from individuals, organizational variables can influence individual willingness and motivation to pursue ideas (Redmond, Mumford, and Teach, 1993).

Also central to our research is the role supervisors[36] play in encouraging employee innovation. Specifically, supervisors can influence subordinate behaviors through mechanisms such as role modeling, goal definition, reward allocation, resource distribution and they can communicate the organizational vision, values and norms and influence subordinates perceptions of the work environment (Bass, 1981; House and Mitchell, 1968; Jaques, 1977; James and James, 1989; Redmond, Mumford, and Teach, 1993; and Yukl, 1986).[37] An open approach to decision making and a considerate management style can influence subordinate creativity (Kanter, 1983; Kimberly and Evanisko, 1981). Supervisors can provide a supportive environment through behaviors that engender empathy, respect, warmth, concreteness, genuineness, trust and flexibility (Isaksen, 1983; Rodgers, 1979; Torrance, 1965). It is asserted that these behaviors are important because they contribute to efficacy beliefs (employee's belief in his/her effectiveness on the job) (Bandura, 1986; Mumford and Gustafson, 1988; Redmond, Mumford, and Teach, 1993). Supervisor behaviors can facilitate creativity through effective goal setting, positive feedback, by managing competing priorities,[38] resource allocations and encouraging problem-solving through training and other skills development (Redmond, Mumford, and Teach, 1993).

In general, supervisor behaviors are important in creating effective organizations (Likert, 1961). And supervisor behaviors that support creativity in individuals are needed if an organization wants employee innovations (Amabile, Conti, Coon, Lazenby and Herron, 1996; Cummings, 1965; Lawler, 1992). Supervisory behaviors that discourage hierarchical barriers, encourage diversity of opinion and openness of subordinate communication, and information sharing across the organization are seen as important (Cummings, 1965; Randolph, 1995). Manager's use of informal feedback as well as formal, dynamic evaluation processes to reinforce the goal of creative solution finding are also considered of key importance (Amabile, 1988; Cummings, 1965).

Also noted in our review of the literature is that the commitment of the supervisor to a particular set of organizational goals can affect employee commitment to perform tasks in that area (Becker, Billings, Eveleth, and Gilbert, 1996). Employees distinguish between commitment to supervisors and commitment to organizations. Commitment to supervisors can "positively and significantly" be associated with performance (Becker, Billings, Eveleth, and Gilbert, 1996; p. 476). If an employee has a propensity to internalize the supervisor's values, then related performance can be predicted (p. 477). Thus, we can predict that manager commitment to environmental protection, environmental innovation and an organizational policy of environmentally sustainable development can have an effect on employee performance in those areas.[39]

Thus, we see there is widespread support in the literature for our research assumption that supervisory behaviors are indeed of paramount importance in communicating organizational encouragement for and personal commitment to such actions, and can positively motivate employee creativity and innovation. Types of behaviors, which can be effectively used to encourage employee innovation, are include in our conceptual framework. (We test whether these behaviors have a statistically significant impact on employee willingness to eco-innovate, as is discussed in Chapter 2.)

Conclusions from the Organizational Behavior Literature: A Model of Factors Affecting Innovation

We looked for an appropriate model of factors that affect employee willingness to innovate, but did not find one in the general organizational behavior literature. But we discovered a model in the organizational learning research of Campbell and Cairns' (1994) which noted the following categories as important for supporting employee learning: Support for Innovation, Training/Capacity, Communication (Participation in Decision-making), Information Dissemination, Use of Rewards and Recognition/Incentives, Sharing of Goals and Responsibilities. These categories of factors parallel our review of the above literature and formed the basis for our development of Figure 1.1. The factors in Figure 1.1 are used to build our conceptual framework (Chapter 2), as well as to compare our findings in the general organizational literature with those we examine in part three of this chapter on environmental management literature.

As Figure 1.1 shows, the factors are related to one another. Some of the relationships we found in the literature are described below. For instance:

- Clarity of goals, managerial attention, information dissemination, participation in the goal-setting decision-making process, and capacity to participate (the set of skills learned that can be applied to the problem), as well as synergies between values involved in goals and ones own values will affect employee participation in innovation.

- Innovation will result more often when goals are ends-oriented rather than means-oriented. And employees will be more likely to innovate if they believe that the organization wants them to search for creative

Figure 1.1   Model of Factors Affecting Employee Willingness to Innovate

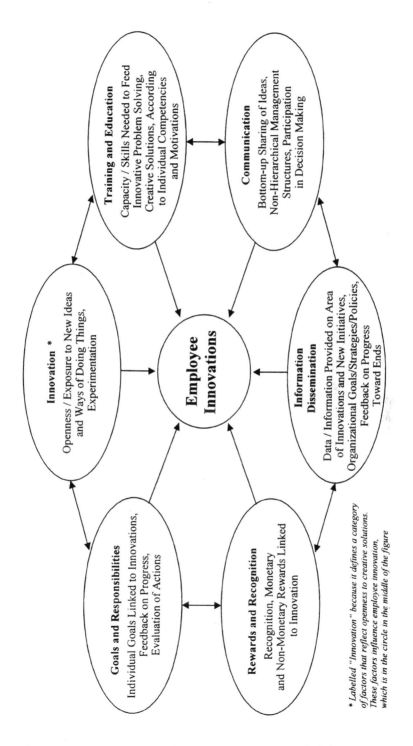

* Labelled "Innovation" because it defines a category
of factors that reflect openness to creative solutions.
These factors influence employee innovation,
which is in the circle in the middle of the figure

alternatives and innovations. Employees will consider a linkage between achievement of innovation goals and rewards/recognition as a positive sign that the organization wants them to focus on those activities.

- Employee capacity to search for alternatives will be limited by the skills and learning the employee has in the search area. It will also be dependent on the skills an employee has in decision making and problem solving, exposure to other ways of thinking and performing actions.

- Measurable and observable goals and sub-tasks with corresponding deadlines to provide time pressure are the best means to focus employee search activities in a desired area.

- Information dissemination helps form the expectations of the employees surrounding the search goals, provide data to help implement innovations, as well as a score-card of progress toward goals. Using these information dissemination tools and networks is an effective means of focusing employee actions in the search area.

- Rewards and other inducements (promotions, bonuses, positive feedback, and formal recognition of achievements) which are linked to initiatives and innovations to achieve specific "ends" are effective motivators for employees.

Part Three: Managing Environmental Change and Eco-Innovation in Firms

Above we have presented the literature findings that organizational and supervisory support are important when a firm wants to encourage employee participation in innovations. From this literature we developed our figure of factors which are important in supporting innovation by employees (Figure 1.1). Here we will discuss the question of whether organizational and supervisory support and these same factors are also key when a firm wishes to support employee environmental innovation. We ask the questions:

- How is environmental management different from general management?

- What organizational behaviors affect employee participation in environmental management? and

- Are the same organizational and supervisory support factors applicable to the case of employee environmental innovations as in the case of other innovation?

What Makes Environmental Management Different From Other Management?

Firms engage in general management in order to deliver a product or service to the market profitably. Whereas, the objective for firms to engage in environmental management can be based on three different factors: reduction of economic impacts, reduction of stakeholder pressures, and/or moral motivation/deep commitment (Winter and Steger, 1997). We discuss these possible objectives below.

In the first case, environmental protection is seen in firms to have positive economic impacts as resources that are not wasted or going out of the company as pollution reduce the bottom-line. Since these resources were originally bought as inputs to production, pollution prevention can result in savings for the company (Thielemann, 1990; p. 45). Taking the economic impact argument one step further, some proactive firms, can also see economic benefits from new product, service, or marketing opportunities from their environmental programs.

In the second case, stakeholders can strongly influence companies to protect the environment, as they put pressure on management to consider the environmental impacts of firm activities. Stakeholders are different groups on which the company depends for its license to operate, like employees,[40] shareholders, society, regulators, etc. (Schaltegger and Sturm, 1991; p. 271). Companies must balance the needs of different stakeholder groups in order to avoid adverse actions they may take against the company. Stakeholders with environmental demands like environmental protection groups, local community groups, consumer advocates, customers, etc. are gaining increasing importance (Meffert, Brenkenstein and Schubert, 1987; Raffee, 1979; Winter and Steger, 1998).[41]

In the third case environmental management is based on the will to protect the planet for future generations and/or improve current conditions for people and nature. Some companies and managers believe that current life-styles and ways of doing business are not sustainable in the long term, and take steps to improve management of activities that have negative environmental impacts (Winter and Steger, 1997). Thus, the commitment and beliefs of the people involved influence the company to protect the environment.

The first two cases cited above for creating environmental management policies and practices (economic and stakeholder/environmental pressure group) are both economic in nature. The third case, that of deep commitment to manage environmental activities is, we believe, the one that differentiates environmental management from general management in firms. In order to be motivated to manage economic impacts, management of a firm must believe that it will make them more profitable to do so. In order to be motivated to manage stakeholder concerns, management of a firm must believe that there will be negative economic consequences in the market place from not addressing these pressures. Both of these types of motivations correspond with the overarching goal of a firm to be profitable. But in the case of commitment, the economic reinforcement for managers to believe in environmental protection is not present. In part two we observed that the values of supervisors could have a direct impact on the conviction of their employees. Below we will discuss how supervisory beliefs about environmental protection can positively influence employee actions. To our mind, this conviction or lack there of on the part of the organization and its members is one aspect of what makes managing the environment different from general management.[42]

We have noted two additional factors that make environmental management different from general management. Environmental management is a relatively new management issue. Unlike management of product quality or human resources, environmental management has begun to be a focus of managerial attention within the last decade or two. Being new means that it is not well integrated into business processes. Second, it is relatively difficult to measure environmental impacts from the business in terms of numbers. (For instance, what is the cost of one ton of air emissions?) Lack of good measurements has led some businesses not to manage these impacts except where mandated by law (Steger, 1998a).

The Importance of Values that Support Environmental Actions in Firms[43]

Environmental behaviors of individuals (including employees and managers) are affected by their values, attitudes, beliefs and knowledge, and understanding these behaviors is the key to finding ways to improve environmental protection (Stern, 1992). Employees' values have special relevance in the case of worker motivation to participate in environmental activities (Rands, 1990). Individual participation in a job assignment is a function of the amount of conflict between the job and the individual's self-image (i.e. individuals have personal preferences which affect their willingness to participate) (March and Simon, 1958). Goitein (1989) reinforces the important role of supervisors' values, noting that the behavior of firms is influenced by the "choice of managers and the presence of individuals who persistently raise environmental concerns within the firm". Environmental protection activities, Rands (1990) argues, must become part of the value-system of managers and employees or the result will be a lack of attention to these types of actions. Kiernan and Levinson (1997) argue that employee satisfaction can be increased by proactive company environmental policies if the employee values environmental protection.

We see that the value an individual places on protection of the natural environment has consequences for the manner in which that employee engages in work related to the environment.[44] In other words, people will behave in a manner which reflects their values in this area. Therefore organizations that want to be sustainable must either hire people who value the environment or find ways of creating environmental commitment from their managers and workers.

To instill commitment in those who do not yet value the natural environment takes a concerted effort by the organization. Shrivastava (1996) states that companies must incorporate environmental considerations into the entire organization by developing a vision, and changing behaviors of line managers. He asserts that the organization must shift managerial values toward coexistence with the natural world. In order to influence managers and employees, he describes the need for organizations to build up core competences through environmental training, sharing of environmental best practices and environmental information, and reinforcing "green" behaviors by employees by using rewards and performance targets which hold the employees accountable for environmental performance in the firm.

Employees may come to a firm with a pre-existing commitment to environmental protection base upon their values. But, Shrivastava (1996) and Stern (1992) argue that it is possible to shift the values of people who do not come with this pre-existing set of values. They argue that by the organization demanding more environmentally aware behaviors from managers and employees, the organization will make employees sensitive to their impact on the natural environment, possibly shifting the values of these individuals in the process.

We conclude that in order to shift an organization and its people from the status quo toward new environmental values and new solutions to environmental problems, management must inspire a change. The impetus for employees to address environmental concerns must come from the organization's vision, since not all employees and managers value environmental protection. As the literature points out, this shift toward more sustainable actions in firms requires organizational, senior management, line management and employee commitment to the change. And, the literature indicates that the same factors that affect employee willingness to innovate (in Figure 1.1) also affect their willingness to engage in environmental change. Below are a number of studies concerning elements that create barriers to and support for environmental change in companies.

What Elements Support Movement Toward Environmental Change in Companies?

Barriers to Environmental Change Post and Altman (1994) write about the barriers to environmental change in organizations. They define two types of barriers: industry and organizational. Industry barriers are uniquely related to the type of business activities of the firm, whereas organizational barriers are not unique to business areas, but rather relate to the firm's capacity to deal with *any type of change*. The organizational barriers that Post and Altman (1994) highlight are

- employee (including workers and managers') attitudes,

- poor communications,

- past practice,

- and, inadequate top management leadership.

Elements that Encourage Environmental Change Case studies by Post and Altman (1994) showed that the effectiveness of environmental champions in organizations depended largely on top management and business unit manager support, as well as expertise and strong appreciation of the environmental problems facing the firm. "The most innovative companies try to create many champions at different levels throughout the organization" (p. 11). Post and Altman (1994) conclude from their research that environmental commitment as demonstrated by environmental policy and strategy statements, changes in organizational structures and corporate culture, alterations in reward and evaluation systems, and adaptation of corporate systems of communication and information sharing are important factors influencing the success of environmental change in organizations. Champions who support environmental protection help the organization to overcome barriers and to share responsibility for environmental policies with employees at all levels in companies.

The environmental studies by Arthur D. Little (1989) and Hunt and Auster (1990) both concur that certain common elements exist in firms that proactively manage the environment. Those elements are

- top level support and commitment,

- existence of and communication of corporate environmental policies,

- daily management systems (oversight and control) that support environmental activities,

- and, sharing ownership of environmental problems with employees.

Stern (1992) adds that general communication, information and feedback, rewards, and evaluation of environmental performance also exist.

Findings from Literature on Barriers and Support for Environmental Change From the above literature we find that top management support /commitment is linked to communication of corporate environmental policies. The greater the commitment of management (as demonstrated by communication of environmental policies) the more likely that employees will have knowledge of the policies. Environmental management literature also indicates that daily management support for environmental change is connected to the degree of employee participation in those change efforts. Finally, the above studies note that behaviors by supervisors can instill employees with ownership for environmental solutions. Furthermore, the factors identified by Arthur D. Little (1989), Hunt and Auster (1990), Shrivastava (1996), Stern (1992) that support the "greening" of organizations by changing environmental behaviors of managers and employees are also those factors that we believe could encourage employee participation in eco-innovation. We test empirically whether policy communication, perceived management support/commitment as demonstrated by managers use of environmental rewards and modification of evaluation systems to include environmental performance and goal setting, open communication and sharing of environmental information affect employee willingness to champion new environmental ideas.

Below we present the literature on environmental commitment looking at overall corporate environmental commitment (CEC), top management commitment, line management commitment and employee commitment. We provide this overview of this literature as the lack of commitment to environmental protection is a major barrier to eco-innovation. And, management commitment to environmental protection can be one of the most important sources of support. Note that this is an area that has been explored by numerous environmental management academics as there is general agreement in the field, that without the types of commitment described environmental impacts of businesses will not be managed, and companies will not become sustainable. Commitment at different levels in firms is therefore a key to eco-innovation in firms.

Management Commitment to Environmental Change

Top management, line management and employee commitment to environmental policies of sustainability can make the difference between success and failure of these policies. Zeffane, Polonsky, and Medley (1995) state that corporate environmental commitment (CEC) is manifest when environmental behavior permeates all levels and activities of the firm. Furthermore, corporate greening requires organizational commitment that extends to company strategy, decision-making, and corporate culture (Carson and Moulden, 1991; Colby, 1991; Friedman, 1991; Garlauskas, 1975; Hunt and Auster, 1990).

Three recent studies have listed criteria that demonstrate the extent of a corporation's environmental commitment. Table 1.1 shows the results of those three studies. In the table we can see that Friedman (1992), Hunt and Auster (1990), and Polonsky, Zeffane, and Medley (1992) agree that support of top management, including senior executives and board members, is important in demonstrating CEC. Friedman (1992) and Polonsky, Zeffane, and Medley (1992) mention that environmental policy is important for showing CEC. Both Friedman (1992) and Hunter and Auster (1990) believe that a key indicator of commitment is the sharing of responsibility with line and functional management. Polonsky, Zeffane, and Medley (1992) do not mention as one of their criteria that line and functional management needs to be responsible for environmental performance, but they do mention it in their 1995 article. These three studies support our assertion that both top/senior management and line/functional management support for environmental actions are important to demonstrate to employees that the firm cares about environmental performance. They also note that environmental policy is

Table 1.1 Criteria Used to Determine Extent of Corporate Environmental Commitment

Hunt and Auster (1990)	*Friedman (1992)*	*Polonsky, Zeffane and Medley (1992)*
• General pro-environmental mindset of managers	• Policy established by the board and senior executives	• Existence of environmental policy and policy implementation
• Resource commitment to environmental activities	• Funds are allocated to ensure effective policy implementation	• Environmental considerations in new investments and ventures
• Top management supports and is involved in the environmental management process	• Responsibility for environmental performance is allocated at the line management level	• Environmental considerations in corporate objectives and performance
• Environmental performance objectives are established	• Environmental objectives are incorporated to all operations and functional areas	• Commitment of board and board members
• Environmental programs are integrated with other programs	• Most advanced procedures, processes, and control methods are used	• Environmental opportunities
• Reporting structures exist in the organization	• Internal compliance system monitors performance	
• Environmental performance reported to top management	• Each division has its own environmental monitoring system	
• Environmental management involves all functional areas	• Environmental information system is implemented	
	• All employees have environmental training	
	• Forward looking environmental attitude in all new areas	

Source: Zeffane, Polonsky, and Medley (1995)

important for signaling CEC, a key assumption underlying our conceptual framework. Below we look at the environmental literature on the role of senior and line/functional management in supporting environmental change.

Top Management Commitment   Rappaport and Dillon (1991) define top management commitment as "involvement and support of top corporate decision makers in environmental matters" (p. 248). This commitment is shown through leadership and direction given by top managers using mechanisms like corporate environmental policies and goal setting which is backed up by making resources (both financial and personnel) available to achieve goals (Rappaport and Dillon, 1991). Davis (1991) states that the

> total commitment by the board and senior managers to a new vision and new values that will create enthusiasm and capture the imagination of the whole work force ... is necessary to start the transformation in a company to sustainable economic development (p. 164).

Stern and Aronson (1984) describe the problem with gaining visible top management commitment to environmental protection as one of scarcity of attention and focus to issues outside businesses daily routines. (A theory that the innovation literature supports, i.e. that focus is often on routines making it difficult to focus energy on innovation in firms (Van de Ven, 1986).) They assert that action is more likely on environmental considerations (energy efficiency, for example) if it is made highly visible within the organization, thus gaining high level managerial attention (Stern and Aronson, 1984). Makower (1993) makes the point that the reluctance of top management to change is often the first obstacle to environmental commitment. And Rappaport and Dillon (1991) state that top management often only shows commitment to environmental matters after an environmental crisis has negatively affected the company.

Nevertheless, Hunt and Auster (1990) state that companies need an environmental mandate from the top to change from past practices and to overcome resistance to environmental change within the organization. But, Maxwell, Rothenberg, Briscoe, and Marcus (1997) state that top management commitment must extend to support of environmental ideas from within the organization. Specifically, "proactive environmental initiatives often come from middle and lower parts of the organization" (p. 132), and top management recognition and encouragement of these initiatives are key to successful environmental strategy implementation (Maxwell, Rothenberg, Briscoe, and Marcus, 1997).

In conclusion, based upon the cited studies, we see that one of the best signals of corporate commitment to employees is a well-communicated corporate environmental policy. As indicated in the literature, a key factor contributing to successful implementation of an environmental strategy in a firm is "visible commitment of senior management to corporate environmental policy" (Maxwell, Rothenberg, Briscoe, and Marcus, 1997, p. 131). Top management commitment to environmental vision is a key impetus for environmental change. And, senior manager recognition of and encouragement of environmental initiatives from middle and lower parts of the organization is also a key to the implementation of a proactive environmental strategy.

Line/Functional Manager Commitment to Environmental Change
Friedman (1992), Hunt and Auster (1990), and Polonsky, Zeffane, and Medley (1992)'s research agrees that line management and functional management commitment are also important in environmental management. Post (1991) notes that environmental problems now affect management of a company's operations, marketing, human resources, finance/accounting and other activities. He asserts that environmental considerations are not peripheral to business decision making but rather, environmental protection is a central commitment of the entire organization (Post, 1991). He goes on to say that beyond the strategic role line managers have in formulating strategies that promote environmental protection, they have an important role in supporting employees. Using reward systems, supporting technical skills development, developing a receptive organizational culture which emphasizes the values associated with environmental protection are all listed as important activities for middle managers (Post, 1991).

Note that this support role of supervisors is particularly important when environmental change is being driven from below. Rasanen, Merilainen and Lovio's (1995) research asserts that informal and bottom-up "greening" of organizations is just as important as top down driven processes. They created a typology of corporate greening processes, defined as processes that improve the sustainability of companies (refer to Table 1.2). Table 1.2 shows that in certain change processes employees in the business units develop the ideas that help the corporation become more environmental proactive, thus making the evolution one that is bottom-up driven. From the table we can see that achieving a policy of sustainability does not only depend upon management, but also rather that employee values and actions can create the impetus for corporate greening. Management support for these employee actions is nevertheless important, as in the absence of management support, we find in our research that these employee actions may be few.

Finally, Wehrmeyer (1995) describes the importance of managerial commitment to environmental performance using a different research perspective.[45]   A study by Wehrmeyer and Parker (1995) shows that managers who have certain behaviors are better able to support corporate environmental performance. These behaviors include:

- interaction with employees as an equal; non-authoritarian or hierarchical,

- open to learning from mistakes; recognizes own ignorance,

- strong leadership; communication of environmental vision/information,

- enables, facilitates rather than monitoring, controlling, and dictating,

- process orientation rather than outcome orientation; learning along the way.

Table 1.2   A Typology of Corporate Greening Processes [a]

Top Down Process of Change	Bottom Up Process of Change
Mission	
Top managers' personal enlightenment and campaigns of cultural change and environmental management	Corporate-wide revolution in the criteria of sustainable business activities
Strategic goal	
Environmental problem solving assignments to the business managers and their staff	Intrapreneuring in green business
Technique	
Environmental problem solving assignments to the specialists	Bootleg projects in the development of environmentally sensitive operations

Source: Rasanen, Merilainen, and Lovio, 1995
[a] Processes that improve the sustainability of companies.

In Wehrmeyer's (1995) view managerial environmental values as demonstrated by observable commitment coupled with a culture of using these behaviors provide better intrinsic motivation to employees resulting in better environmental performance.

For our study, we note that not only are managerial values and commitment important but that scholars agree that daily behaviors that support employee participation in environmental idea creation are very significant. Line management behaviors which support employees include use of rewards systems, support for environmental competence building, openness to employee participation using non-authoritarian style of communication and widespread sharing of pertinent information. Thus, just as the general organizational behavior literature reviewed in part two of this chapter highlighted the factors necessary to support innovation, we find that the environmental management literature on corporate commitment finds the same factors to be of importance in supporting environmental change. Behavior change is noted in both literatures as a key to this support process.[46]

Employee Environmental Innovation Literature Supporting Choice of Factors

The previously cited environmental management literature was concerned with showing that environmental change toward sustainable development needs to be supported by commitment of different groups of managers at different levels in the company. The literature suggested that certain support factors demonstrate the level of commitment or signal of commitment for environmental protection and sustainable actions. Here we look specifically at the literature on employee eco-innovation. (Note that the above literature is completely different than the authors cited below.) We cite the literature below to demonstrate that the suppositions that we drew from the environmental change literature are indeed supported in the specific literature on employee eco-innovation. (Said another way, the above literature is focused on organizational commitment to sustainable actions, and the literature below demonstrates the types of support employees need to become involved in eco-innovations, the subject of our research.)

In our research, we are concerned not only with mechanisms and programs that can increase employee participation in environmental policy in general, but more specifically those organizational supports necessary for environmental innovation. Hostager, Neil, Decker and Lorentz's (1998) and Keogh and Polonsky (1998) research shows that organizational support is necessary to encourage environmental innovations.[47] Lefebvre, Lefebvre, and Roy (1995) show that integration of environmental issues into corporate strategy can be a catalyst for radical organizational innovation. Involving managers and stakeholder at all levels in this strategy, including top management, operations head, shareholders, research and development managers can lead to a more successful outcome (Lefebvre, Lefebvre, and Roy, 1995).

The types of support that employees need to become involved in environmental innovations were looked at by Milliman and Clair (1995) who surveyed organizations with innovative environmental practices. They found that human resource programs that were designed to facilitate employee attainment of environmental vision and innovation included:

- training on environmental skills and knowledge,

- performance appraisals that make employees accountable for environmental performance, and,

- rewards that provide incentives to accomplish environmental objectives.[48]

These programs helped to increase employee participation in environmental innovations in the firms (Milliman and Clair, 1995). Storen (1997) specifically shows how information systems, performance measures and targets and inspiration/knowledge from education feed the human processes necessary to develop relevant eco-innovations.[49] Thus, we can see that factors of environmental training, performance appraisals that make employees accountable for environmental performance, rewards that provide incentives to accomplish environmental objectives, environmental information systems, and inspiration/support for eco-innovation are all factors that the literature indicates, using case study evidence, that are important.

Hostager, Neil, Decker and Lorentz's (1998) research indicates that innovation in the environmental area depends upon the firm's ability to support individuals and groups in those tasks.[50] To look at the issue of what motivates employees to participate, specifically, in environmental innovations, Hostager, Neil, Decker and Lorentz (1998) propose a model which builds on the tenet that performance is a function of ability and motivation (Bird, 1989; Mitchell, 1978). They defined ability as individual capability and organizational resources, and motivation in terms of societal motivators like cleaner environment, and individual motivators, including extrinsic (salary, promotion, status) and intrinsic (pride, accomplishment, challenge) motivators. The authors add the self-efficacy component and desirability, which is defined as attitudes, social norms, intrinsic and extrinsic rewards, and punishments and disincentives which signal the desire of the firm for the individuals to eco-innovate (ibid, p. 17).

Thus, employees come to a firm with their own capability and intrinsic motivation, which affects the likelihood of their trying to promote an eco-innovation, but the firm itself provides resources and extrinsic motivations like desirability and organizational incentives that can either encourage or discourage environmental initiatives.[51]

1.3 CONCLUSIONS FROM LITERATURE REVIEW

Repeatedly the literature indicates the need for proactive environmental policies in order for a firm to become sustainable in its activities. It also emphasizes the need to encourage supervisory behaviors to support employee involvement in environmental change and innovation. These factors are the same as we found to be of importance in supporting innovation in general (see Figure 1.1). But we find an emphasis in the environmental literature on not only the use of these factors but a need for corporate commitment to sustainable actions. The literature states that corporate focus on the environmental innovation goals of the organization using these factors to encourage individual action in the environmental area is of paramount importance if companies want to achieve ecologically sustainable development.

Thus, we see that in the environmental case, commitment to environmental change has a special significance. The organizational greening literature indicates that employee and management values regarding protection of the natural environmental can be positively shaped by a mandated commitment to a proactive environmental policy. Thus, a company must present a vision and incentives for employees at all levels in the company to follow that vision. Senior management can positively affect organizational commitment, by demonstrating visible support for and communication of the environmental policy. Supervisory personnel can support this vision by using daily support behaviors to encourage employees to apply their skills to finding environmental solutions.

Specifically, the literature points to the importance of line management support for learning as demonstrated by behaviors that encourage:

- innovation/experimentation/learning from mistakes/sharing knowledge in the environmental area,

- education to develop skills and ideas to address environmental matters,

- open communication/lack of hierarchies/participation in environmental decision-making,

- use of information sharing to inform employees of environmental vision and strategic direction,

- use of formal rewards/incentives and informal recognition to motivate environmental participation,

- and sharing of environmental performance objectives and responsibilities.

In the following chapter we will present our conceptual framework, built upon the literature review and assumptions which we have presented above.

Chapter 2

Conceptual Framework for Empirical Investigation

In Chapter 1 we derived from the literature an understanding of the possible importance of organizational and supervisory encouragement factors in employee environmental innovation. We have developed a framework to test which of these factors are perceived by employees to support their environmental initiatives in companies. In this chapter we present our conceptual framework for the empirical study, building it out of the factors we identified in Figure 1.1.

We have taken the following approach in this chapter. In the first part of this chapter we build the conceptual framework and then in the subsequent parts of this chapter we give a detailed description of how each component in the framework was developed and defined (explaining why the components were chosen). For example, in the first part we present the fact that the conceptual framework has eight hypotheses (one for environmental policies, six for the factors identified in Figure 1.1, and one which compares supervisor support for environmental versus other general management activities). Then in the following parts of the chapter we systematically proceed by describing how we decided what would be tested in each of these hypotheses.

Thus, in part one we outline the organizational and supervisory support components of the framework and how this fits into the overall natural environment. In the second part we describe how we used literature on environmental management systems and international environmental charters to select our list of thirteen environmental policies from which we measure organizational encouragement in the survey. Also in part two we define each of those policies. In the third part we define the six behavioral categories that form the hypotheses used in the empirical study to measure the employee perceptions of the signals of supervisory support for eco-initiatives. (See Figure 1.1 for the categories.) Finally, we describe the last hypothesis which is a comparison of the data from the supervisory behavior variables.

2.1   PART ONE: CONCEPTUAL MODEL FOR EMPIRICAL INVESTIGATION

Figure 2.1 depicts the conceptual model underlying our survey instrument. This figure notes that regulatory and stakeholder pressures create an impetus for environmental actions in companies, resulting in environmentally proactive firms sending signals of organizational and supervisory encouragement to employees.   It indicates that if the respondents perceive a strong signal from the causal variables of the model (organizational encouragement and signal of supervisor encouragement) than this will positively influence employee self-described environmental initiatives, which will improve company environmental performance. (The strength of the signal is determined by the degree of positive (or negative) perceptions of employees. Either their perception of the ranked behaviors which are listed from least to most positive in the survey, or their perception of the strength of company commitment to environmental policies.) The relationship between employee perceptions of support and employee self-described environmental initiatives is tested using eight hypotheses. one for the organizational signal and six for the supervisor signal. And in the last hypothesis, we measure whether supervisory behaviors that support environmental management versus behaviors that support other management activities have the same effect on the dependent variable. The sources of the causal variables of organizational and supervisory encouragement, and the categories of factors that form the independent variable, were discussed above in Chapter 1 literature review.

It is our first general prediction that a strong signal of organizational support from environmental policies will positively affect employee willingness to eco-innovate. Environmental policy (broadly defined to include corporate vision and strategy as by Brophy, 1996) has been shown to be an important precursor to employee engagement in environmental activities (Barrett and Murphy, 1996; Brophy, 1996; Hutchinson, 1996). Therefore, we use as a proxy for organizational encouragement, employee knowledge of the existence/perceptions of organizational commitment to the environmental policy. We first test whether, if employees perceive a strong signal from the organizational environmental policies, this will positively affect their willingness to eco-innovate (*Hypothesis 1*). Within the environmental policy area we asked about the employee's perception of organizational encouragement from thirteen different environmental policies. (These policies are defined below.) The strength of the signal depends upon both the formulation of such policies and the degree to which they have been clearly communicated to employees.[1] It is assumed that the clearer and stronger the employees' perceptions of these policies, the more

Figure 2.1 Conceptual Model for Assessment of Employee Perception of Organizational and Supervisory Encouragement of Environmental Initiatives

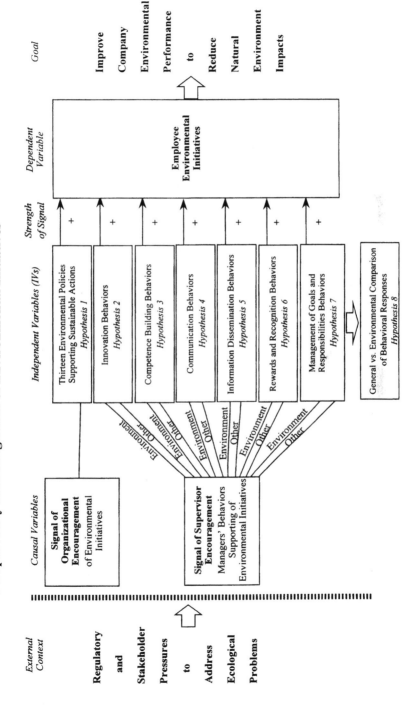

likely that they will engage in environmentally proactive initiatives or innovations which create value for the company.

Of these thirteen policies, the first item asked about is employee knowledge of the company's published environmental policy. The other twelve policies, which in most cases are included in this general policy statement but may be written in separate policy documents, are subsets of this general policy. They include specific policies like purchasing, fossil fuel use reduction, etc. It is likely that not all excellent practice companies will have all of the environmental policies discussed below. It is assumed that just because a policy exists in a company in written form, it does not necessarily follow that the policy will be clearly communicated to employees, nor that employees will believe the management within the company is committed to that policy.[2] All of these policies are given equal weight in our analysis of the data.[3]

Our second general prediction is that employees who perceive that their supervisors are supportive, as demonstrated by daily observable behaviors, will be positively influenced to try environmental initiatives. To test employee perceptions of supervisory encouragement we developed seven hypotheses.

These hypotheses relate to six categories defined in Chapter 1, Figure 1.1 as those areas that affect the willingness of employees to innovate and which we determined from the literature to also be the areas that affect employee willingness to *eco*-innovate. Campbell and Cairns (1994) indicate the importance of the same six areas of Innovation, Competence Building, Communication, Information Dissemination, Rewards & Recognition, and Measurable Goals and Responsibilities for organizational learning. The literature on organizational behavior and innovation indicates that supervisory support behaviors in these categories are important for employee initiatives. And the environmental management literature also indicates the importance of behaviors in these categories in influencing employee environmental participation and innovation. But no empirical study has previously tested the relative importance of these factors influencing employees' eco-initiatives. And, in addition to determining which of these factors have a statistically significant impact on employee eco-innovation, our survey tests whether the same supervisory behaviors which are believed to support general organizational learning are also those which influence employees' environmental innovations.

Specifically, supervisory behaviors in these six areas are individually tested to assess their influence on employee environmental initiatives (*Hypotheses 2-7*). Note that the categories and the behaviors are both related to general management. In the survey, we asked employees to select a supervisory behavior in each category which a) is typical in the general case and b) is typical when their supervisor is managing environmental issues. This allowed us to compare supervisory support behaviors in general management versus environmental management. We hypothesize (*Hypothesis 8*) that when applied to the environmental case, these behaviors will have a stronger and more significant impact on employee willingness to try environmental initiatives than when applied to general management. This allows us to draw conclusions about different behaviors that supervisors are using when managing environmental versus other management activities.[4]

2.2   PART TWO: IMPORTANCE OF ENVIRONMENTAL POLICIES TO SIGNAL ORGANIZATIONAL ENCOURAGEMENT

There is general agreement in the literature as to the importance of environmental policies as a signal of organizational commitment to and encouragement of environmental activities within companies (Arthur D. Little (1989), Hunt and Auster (1990), Maxwell, Rothenberg, Briscoe, and Marcus (1997), Polonsky, Zeffane, and Medley (1992), Post and Altman (1994), Rappaport and Dillon (1991) in Chapter 1). We test empirically whether this untested theory is indeed reasonable.[5]

Environmental Policies: An Analysis of Management Systems and International Charters

There are two possible frameworks for determining which environmental policies could influence employee eco-innovations in firms: environmental management systems (including life cycle analysis[6]) and international environmental charters. We review both of the possible frameworks in this section.   The sources for our analysis of policies used to implement environmental management systems and life cycle analysis are the International Standard Organization's environmental management standard (ISO 14001) and the European Union's Environmental Management and Auditing Scheme (EMAS). [7,8]   For life cycle policies, and to verify that these policies actually exist in European companies, we use the results of the longitudinal European Environmental Management Survey (called Barometers). Second, we looked at environmental charters and principles developed by non-governmental organizations over the last decade.   These charters give another type of blueprint for environmental policies of sustainable development in firms, indicating what industry-specific and non-industry-specific environmental charter organizations expect from companies. The environmental management systems and charters revealed an overlapping set of policies which became the basis for our survey list.

Environmental Management Systems and Life Cycle Analysis

Environmental policy development in firms is one component of standardized environmental management systems, like EMAS and ISO 14001. The implementation in a company of one of these management systems is a necessary, but not a sufficient condition for reaching ecological sustainability of activities (Gray, Bebbington, and Walters, 1993; Roome, 1992). Welford (1993) states that an environmental management system is sub-optimal if the ultimate objective is sustainable development, but that these systems help personnel to understand their responsibilities and the structure of environmental management in the company. Welford (1993) argues that linking environmental management systems with Life Cycle Assessment/Analysis (LCA) is an effective tool for developing a system which can help a company become ecologically sustainable. Thus, as we see below, both environmental management systems and life cycle assessment are essential components of proactive company policy development.

According to Netherwood (1996) standard policy components of an environmental management system are the following (this list is based on his analysis of ISO 14001 and EMAS):

- organizational commitment to environmental management,

- adoption of an environmental policy,

- implementation of environmental training,

- allocation of environmental responsibility,

- setting of environmental objectives and targets,

- reporting of environmental performance.

But, Netherwood's list of policies that exist in companies with environment management systems is not complete. It does not include components of life cycle assessment, which are thought to be necessary to a more proactive policy, especially in the area of product development (Starkey 1996; Welford 1993).[9]

In order to make the list more comprehensive; we have identified a study that confirms the existence of EMS and LCA policies within European companies. The European publication entitled *International Business Environmental Barometer 1997* (Belz and Strannegard, 1997) reports the results of a bi-annual longitudinal survey which measures the status of environmental management across countries over time. The International Business Environmental Barometer developed their survey using as inputs the European Union Environmental Management and Auditing Scheme (EMAS) and ISO 14001 and components of life cycle assessment.

The Barometer survey effort began in 1993. The latest published results (1996) involved eight academics from five academic institutions (one in each country). The survey resulted in a random sample of 607 businesses in Belgium, Norway, Sweden, Switzerland and Finland, with response rates from between 25 and 45 percent. (Note that the survey is being conducted again in 1997 with companies in twelve European countries, but those results are not yet available.)

The Barometer questionnaire included questions on implementation of environmental management systems, i.e. whether the organization had

- a written environmental policy,

- environmental reviewing (auditing),

- measurable environmental objectives for continuous improvement,

- an environmental program (or management system),

- sharing of environmental responsibility,

- an environmental report, and

- environmental information in annual report[10] (Belz and Strannegard, 1997).

In addition the survey asked a series of questions referring to the life cycle impact of the product in order to assess company movement toward eco-efficiency and sustainability. The survey asked about organizational policies to:

- reduce raw material and fuel/utility use through out the product life cycle,

- reduce usage/disposal of toxic substances,

- reduce solid waste/conserve materials (water, raw materials, and soil),

- reduce emissions,

- save energy (use renewable sources of energy),

- include environmental considerations into other company policies, including production, recycling/disposal, procurement, logistics, research and development and marketing/sales, and

- closed loop production systems (Belz and Strannegard, 1997).

From above policy components of EMS and LCA we developed an initial list of policies that indicated organizational commitment to sustainable development. (Please refer to the first column of Table 2.1 for list derived from research on environmental management systems and life cycle analysis.)

International Environmental Charters

Our second approach to developing a list of environmental policies that exist in proactive firms was to look at the policies companies commit to implement when they sign on to one or more of the international environmental charters. By examining the components of these charters we saw that similar policies were indicated to those identified above.

International charters are documents, which companies sign as an indication of their intent to follow specific environmental principles aimed at helping businesses manage and improve their environmental performance (Brophy, 1996). According to Datta (1995), eighty percent of Fortune 500 firms have written an environmental charter with many signing CERES or other industry charters. Many groups developed charters. The International Chamber of Commerce and Environment Business Forum, an initiative of the Confederation of British Industry (in 1991) developed the Business Charter for Sustainable Development. The Business Council on National Issues, the National Roundtable on the Environment and the Economy (NRTEE), Keidanren, Chemical Industry Association (CIA),

Canadian Chemical Producers' Association (CCPA) and the European Petroleum Industry Association (EUROPIA) also have charters and guidelines. According to Brophy's (1996) thorough examination of the above organizations' charters, CERES (Coalition of Environmentally Responsible Economies) is the most complete of all of these charters. It is also unique because it was developed by non-industry affiliated groups.

CERES Principles were developed by a group of United States environmental groups and members of the social investment community after the Exxon Valdez oil spill disaster in April 1989 in order to prevent future disasters. In the introduction of the CERES Principles there is a statement that "corporations must not compromise the ability of future generations to sustain themselves" (CERES, 1995). The categories covered include the following:

- Protection of biosphere - reduce or eliminate the release of harmful substances.

- Sustainable use of natural resources - use and conservation of renewable resources (water, soil, forests, etc.).

- Reduction and disposal of waste.

- Energy conservation - use and conserve safe, sustainable energy sources.

- Risk reduction - minimize environmental risks and be prepared for emergencies.

- Safe products and services - reduce or eliminate use, manufacture and sale of products and services that cause environmental damage.

- Environmental restoration.

- Informing the public - when company causes conditions that endanger environment and human health.

- Management commitment - of board level and Chief Executive Officer.

- Audits and reports - annual public report of environmental performance.

CERES fails to mention target setting/objectives, raising employee and supplier awareness, life cycle assessment, and environmental impact assessment, all of which are mentioned in some of the above mentioned industry organization charters (Brophy, 1996). (Refer to the second column of Table 2.1 for policies included in the international charters.)

Analysis of Overlap between EMS/LCA and International Charters

From the environmental management standards, life cycle assessment and international charters we have developed a complete picture of what proactive environmental policies might exist in a company. We have created the following table to compare EMS and LCA policies (tested for in the Barometer survey) with the policies mandated in the international nine international charters examined above.

Missing from the Charters list were the explicit mention of a written environmental policy and of an environmental management system (items number 1 and 4). These two policies exist in proactive companies and are required by the environmental management standards.[11] Missing from the EMS/LCA list was the explicit mention of reducing use of unsustainable products or, said another way, the use of renewable resources (item number 12). But the Barometer Survey mentions reduced materials use and conservation repeatedly and indicates that life cycle impacts are of paramount importance. Thus, while this item is not explicitly mentioned, it is implicit in their Barometer questionnaire. We have included it on our list.[12]

Development of Our List of Thirteen Environmental Policies

When creating our final list of policies, we considered the input from the above sources, but modified the inputs to some degree. Beginning with the above list of twelve policies, we added a policy, which exists in some companies, concerning the application of equally stringent environmental standards at home and abroad. This question was particularly relevant as the companies in our research sample were large corporations which have impacts across national boundaries, and as this is a differentiating policy of which the employee perception can indicate a great deal about how strongly they feel their management is committed to environmental protection. Missing from our list of policies is an explicit question about the company's recycling/waste reduction policy since this question is embedded in the measurable target question.[13]

Table 2.1   Comparison of Policies in EMS/LCA vs. International Charters

	EMS + LCA		International Charters[a]
1	Written environmental policy	1	(Implicit- see below)[c]
2	Measurable environmental objectives	2	Target setting/Environmental objectives
3a	Environmental review/Audit system	3a+b	Audits/Reports - annual public reports of environmental performance
3b	Environmental report[b]/Environmental information in annual report		
4	Environmental programs/Management system	4	(Implicit- see below)[c]
5	Environment as part of procurement policy	5	Raising supplier awareness/Safe products – reduce/eliminate use, manufacture & sale of products that cause environmental damage
6	Environmental training (part of both EMAS and ISO 14001)	6/7	Raising employee awareness
7	Sharing environmental responsibility throughout organization	7	Above (part of employee awareness and sharing of targets)
8	Use life cycle analysis to reduce materials use/Solid waste/Conserve materials/Use closed loop systems and extend product life cycle	8	Use life cycle/Environmental impact assessment/Reduction and disposal of waste/Reduce/Eliminate use, manufacture and sale of products that cause environmental damage
9	Eco-efficiency/Sustainability criteria	9	Sustain the environment for future generations/Management commitment at board level and chief executive officer through organization
10	Reduce fuel use/Save energy	10	Energy conservation – use safe, sustainable energy sources and conserve energy
11	Reduce toxic chemical use/reduce emissions and risk	11	Protect biosphere by reducing resource use and eliminating release of harmful substances
12	Conserve materials	12	Sustainable use of natural resources – use renewable resources and conserve

[a] CERES plus five additional items mentioned in other industry group charters.
[b] Included in EMAS but not ISO 14001 standard.
[c] When charters were formulated EMSs were not an issue. They were developed to implement environmental management objectives. (According to U. Steger, Professor of Environmental Management at IMD International, December 1998.)

Based on the above analysis, we have developed a list of policies which we tested for in the survey. (Please refer to Table 2.2.)

As we previously mentioned, we kept the list of questions short and the language simple, so that employees from different cultures, levels of sophistication and education could easily understand it. Companies working with us to conduct the survey insisted that the survey be kept as concise as possible. Thus, several items are imbedded in some of the questions. A water utilities company in the United Kingdom, Northumbrian Water Ltd., tested the survey. Then the policy questions were reformulated using feedback from the survey test and from a focus group with environmental managers. (Focus groups and questionnaire development are discussed in Chapter 3 on methodology.)

Table 2.2   Corporate Environmental Policies

My company...
1. publishes an environmental policy.
2. has specific targets for environmental performance.
3. publishes an annual environmental report.
4. uses an environmental management system.
5. applies environmental considerations to purchasing decisions.
6. provides employee environmental training.
7. makes employees responsible for company environmental performance.
8. uses life cycle analysis.
9. has management which understands/addresses issue of sustainable development.
10. systematically reduces fossil fuel use.
11. systematically reduces toxic chemicals use.
12. systematically reduces consumption of unsustainable products.
13. applies the same environmental standards at home and abroad.

Definition of Environmental Policies

Each of the thirteen items listed in Table 2.2 are defined below.

Published Environmental Policy

1. My company publishes an environmental policy.

An environmental policy forms the backbone and skeletal framework from which all other environmental components are hung (including environmental management systems, audits, assessments, and reports) (Brophy, 1996). An environmental policy can indicate to internal and external stakeholders that the company intends to take environmental issues seriously. As such it is an information and communication tool. It serves an informative role of notifying bankers, insurers, and shareholders of its intention to protect the natural environment. It tells employees that the company will take a responsible approach to environmental issues. And, it can allay fears of community and environmental groups about the organization's intended approach to environmental affairs (Brophy, 1996). In addition, policy can serve as a functional tool, acting as a guide for employee actions in that it usually includes environmental targets and objectives, setting parameters and boundaries within which environmental actions can be taken.

Research by Brophy (1996) indicates that having an environmental policy does not in itself make a firm proactive in the environmental area. Rather, it is a minimum prerequisite for a company who wants to move toward environmental value creation and to begin including employees at different levels of the company in environmental responsibility and performance. Epstein (1996) agrees with this point, stating that a clearly communicated environmental mission statement can demonstrate top management commitment to a proactive environmental strategy, but it may be insufficient if the goal is integration of environmental concerns into all company decision-making. Thus, in our survey our starting point was employee knowledge of the existence of a written environmental policy, but we went on to ask the employees about their perceptions of the existence of twelve other environmental policy components.

We define environmental policy as a written document that expresses an organization's intentions with regards to protecting the natural environment, including in some cases organizational targets and goals. The environmental policy of a company may include a complete blueprint of the environmental program of the firm (including strategy, means of attaining that strategy, and other sub-policies) or it may be a simple document which states that the company will take a proactive approach to manage its environmental impacts. (Note that we are only concerned in our empirical study with companies who intend to take a proactive approach to environmental protection.)[14]

Targets for Improving Environmental Performance

2. My company has specific targets for improving its environmental performance.

Epstein (1996) emphasizes the need for systems to measure environmental impacts across the life cycle of a company's products and services. Once these impacts are quantified, he suggests that the manner in which a company can drive environmental performance improvements is by using an environmental performance index which lays out targets for facility, business unit, and corporate performance (Epstein, 1996). Due to the predominance of the use of performance measures and targets as a means for managing environmental impacts, we have included a survey question asking employees about their knowledge and perception of company commitment to use of performance targets.

We define this policy as the use of performance measures and targets[15] in order to focus organizational energy on improving environmental performance. The environmental targets can be set for the entire organization, for individual business units, and/or for individuals/groups of individuals. They can include resource use reduction goals, product redesign goals, recycling goals, and/or any of a number of other targets aimed at reducing the environmental impacts of business operations over the entire life cycle of the organization's product/service.

Publishes a Meaningful Annual Environmental Report

3. My company publishes a meaningful annual environmental report.

According to Sustainability and United Nations Environment Program 1997 report, companies with a demonstrated commitment to environmental management have increasingly, in the last five years, begun to publish their performance for stakeholder review. There are two reasons for environmental reporting, communication to external stakeholders and communication to employees. For our research the more important reason is the use proactive companies make of environmental reporting as a tool for internal employee communication and motivation. Many companies state that the external audience is of secondary concern for them when publishing their report. Rather they believed that the report is an important tool in the annual environmental goals setting process, and many believe that the transparency of reporting of performance on the goals helps to motivate involvement of employees throughout the company. Lober, Bynum, Campbell and Jacques (1997) state that reporting "can be a tool for increasing employee involvement in environmental management, increasing employee morale, and winning top management support" (p. 73).

We define environmental reporting as the publishing of a written document, which accurately presents the environmental objectives of the organizations and annual or biannual performance toward explicit targets. This document should be widely available within the organization and to external stakeholders. For our research we are interested in how this document is perceived by employees in the organization, therefore we ask whether or not the employee finds the document meaningful.

Introduced an Environmental Management System

4. My company has introduced an environmental management system.

Above we have discussed the international standards for environmental management systems in some detail. These systems help companies to implement environmental policy and environmental performance improvements.

We define environmental management systems (EMS) as a management tool for integrating environmental protection policies into business processes. An EMS uses a systematic approach to enabling people/working groups in the organization to take environmental impacts into consideration when performing daily job functions, and helps the organization to achieve its stated environmental targets and objectives.

Introduced Environmental Considerations into its Purchasing Decisions

5. My company has introduced environmental considerations into purchasing decisions.

Brophy (1996) notes that a standard part of many companies' environmental policies is a statement that the company will work with suppliers to minimize impacts on the environment. Epstein (1996) states it a different way, claiming that companies can "put pressure on suppliers to reduce the negative environmental impacts of the components of the products and services they provide" (p. 46). Whether it is working through a partnership with suppliers or applying pressure, a company can positively affect the environmental impact over the life cycle of their products and services by introducing environmental considerations into its purchasing decisions (Green, Morton, and New, 1995).

The definition of environmental purchasing we use in our survey is a decision-making framework that encourages the buying of components or services[16] which were produced or delivered with reduced environmental impact compared to alternatives and/or which improve the environmental performance of the company's product or service. Purchasing, in our definition, can occur at any stage in the life cycle of the company's business operations, including any item or service from inputs to production to office paper and supplies.

Employees Training Regarding Environmental Issues

6. My company provides employees with training regarding environmental issues.

The environmental management literature of authors like Ulhoi, Madsen, and Rikhardsson (1994), Wehrmeyer (1996), Welford (1996), and Winter (1988) states that environmental training, education and skills development play an important role in harnessing employee enthusiasm for environmental change and improving employee ability to give high quality contributions to environmental activities. (Note that the organizational behavior literature reviewed in Chapter 1 also emphasizes the role of education in motivating employees.) Therefore, we have included a question asking whether employees agree that their company provided them with training and education on environmental issues.

Training on environmental issues is broadly defined in our survey as any training, education or skill development activity that the employee perceived to be available to employees in the company. Note that what we were interested in ascertaining from this question was whether the employee perceived that the company actually provided employees with the training. If the training existed, but the company and its managers were not encouraging employees to attend then we would expect that the perception would be negative. (In our definition we did not judge the quality of the training.)

Employee Responsibility for Company Environmental Performance

7. My company makes employees from all levels of business responsible for company environmental performance.

Brophy (1996), Hutchinson (1996), Wehrmeyer (1996) and Winter (1988) all assert that for environmental policy and strategy to be implemented, environmental responsibility must be incorporated into all levels of the company. "Until environmental considerations are applied to every aspect of an organization's functioning, and every employee is aware of their responsibilities, an environmental policy can exist in name only" (Brophy, 1996; p. 94). (Note that the organizational behavior literature in Chapter 1 states that employee responsibility is important for engaging employees in business operations and in creative solution finding.)

Thus, we ask the employee for his/her perception of whether the company makes employees at all levels of the company responsible for the environmental performance. The definition we use for environmental responsibility in our survey is entirely based upon the respondent's perception.[17] We are asking for respondent perceptions about whether responsibility for the environmental impacts of the business reside with managers alone, or with the employees, including themselves.

Life Cycle Analysis Policy

8. My company uses life cycle analysis as a means of assessing and minimizing the environmental impacts of its processes or products.

LCA is a tool for designing for reduced environmental impacts of products, but also, technologies, materials, processes, industrial systems, activities, and services (SETAC News, 1993). Welford (1996) notes that LCA can be an effective tool for aiding companies who want to "achieve real sustainable development" (p. 215).

In our survey we do not define life cycle analysis for the respondents, beyond asking them directly if the company uses this tool for assessing and minimizing the environmental impacts of company business processes or products. We assume that the company has either communicated the usefulness of this tool and its integration into company decision-making or has not.

Management Understands Sustainable Development

9. My company's management understands the wider issues of sustainable development and is beginning to address them.

There is an active movement within excellent practice corporations today to take responsibility for sustainable development of their enterprises as can be seen most notably in the work of companies involved in the

World Business Council for Sustainable Development (WBCSD, 1996a,b,c). This trend in corporate practice has been substantiated as a necessary step toward sustainable development in the academic literature. (Refer to the environmental literature in Chapter 1. For example, Ayres (1995) and Gladwin (1992), who both highlight the role of corporate managers in bringing about the transformation toward environmentally sustainable businesses. And, Frankel (1995) who asserts that to become truly sustainable in activities, corporate executives must play a central role in bridging the gap between traditional business approaches and those necessary for the environmental transformation into a sustainable enterprise.)

Since our research in proactive companies is concerned with how employees perceive management support for environmental initiatives, a question on management support for sustainable development was relevant to our study. Therefore, we have included a question in our survey asking employees if they perceive that their management understands the issues of sustainable development and is addressing them.

Note that this is a general question to see if employees perceived management support for sustainable development to be part of the environmental policy of the company. The BARS under hypotheses 2-8 are concerned with specific behaviors of supervisors, which can either demonstrate commitment to environmental activities (or fail to do so). Note that this general policy measure is concerned with perceptions of overall organizational commitment of managers, and the specific BARS are concerned with the behavior of the respondent's direct supervisor.

Four Questions on Company Sustainability Policies

10. My company plans to systematically reduce its dependence on fossil fuels.

11. My company plans to systematically reduce its use (or manufacture) of toxic chemicals.

12. My company plans to systematically reduce its consumption of unsustainable products.

13. My company applies the same high environmental standards to its activities at home and abroad.

The four policies listed above have a common foundation in the sustainability frameworks of The Natural Step (described in detail in Chapter 1) and Integrated Chain Management, which we will describe below. Integrated Chain Management (ICM) provides a framework for companies to systematically reduce fossil fuel energy sources, replacing them with renewable/sustainable sources; preventing pollution by finding substitutes for (reducing use of) environmentally-hazardous substances such as toxic chemicals; and conducting life cycle assessments across the company value chain to find opportunities for environmental improvements (Cramer, 1996a). Companies have developed policies to incorporate the use reduction principles and environmental equity conditions that are highlighted in The Natural Step and Integrated Chain Management.

We define the three first policies to be a systematic (methodical) reduction of the use/manufacture of non-sustainable energy, chemicals or other products. We are interested in whether employees believe that the company is implementing a policy to reduce use of these three categories of resources. For instance, fossil fuels are a non-renewable source of energy which companies can reduce consumption of by reducing overall energy consumption or switching to renewable energy sources. Toxic chemicals are defined as persistent chemicals that create damage to the natural environment for which companies can find substitutes. Unsustainable products are goods that are taken but not replaced in the natural environment (like rainforest timber), thus depleting the natural ability of the planet or a particular ecosystem to sustain its long term health.

The fourth policy of environmental equity is asking whether employees believe that the company uses the same environmental policies in its worldwide operations to protect the natural environment and employees/people who exist in other geographical locations. (Refer to The Natural Step's fourth system condition for the principles underlying this policy question.)

2.3　PART THREE: SUPERVISORY BEHAVIORS THAT SUPPORT EMPLOYEE INITIATIVES

Above, in part two of Chapter 2, we have discussed our process for selection and definitions of the environmental policy factors underlying the first hypothesis in the conceptual model. In part three of Chapter 2, we define the six categories of supervisory behaviors that we determine might influence employee willingness to eco-innovate. These factors form the basis for testing hypotheses 2-8 in the conceptual model. Please note that from the Chapter 1 literature review we have already determined the importance of supervisory support to employee willingness to innovate and we have defined the six categories in which supervisory support might be expected to influence employee innovation. (Please refer to Figure 1.1 for the six categories of factors defined.)

Definition of Supervisory Behavior Categories

As noted in our description of the conceptual framework at the beginning of this chapter (Figure 2.1), supervisory behaviors that support employee initiatives fit into six categories: Innovation, Competence Building, Communication, Information Dissemination, Rewards & Recognition, and Measurable Goals and Responsibilities. The six categories are defined below.[18]

Innovation

Hage and Dewar (1973) have demonstrated that the values[19] of supervisors can impact the rate of innovation in organizations. Other studies have demonstrated the role of encouragement of risk-taking, idea generation and valuing innovation from all levels of the organization (Cummings, 1965; Delbecq and Mills, 1985; Ettlie, 1983; Hage and Dewar, 1973; Kanter, 1983; Kimberley and Evanisko, 1981). And, Burgelman (1983) highlights encouragement of experimentation as positively influencing business innovation. Within the area of management support for innovation, *Hypothesis 2*, we test behaviors by supervisors that have been demonstrated in the literature and in interviews to have a positive or negative effect on creation of new ideas and promotion of innovations within a company. Included in this category are rank-ordered behaviors ranging from "usually objects to changes and new ideas" to "encourages partnerships with other departments in order to implement new ideas".

Competence Building

Support in organizations for the attainment of expertise positively influences employee creativity (Langley and Jones, 1988). Ulhoi, Madsen and Rikhardsson (1994) and Wehrmeyer (1996) have studied the effects of employee training on their participation in environmental activities, highlighting the importance of this area in encouragement of environmental innovations. In the area of management support for competence building, *Hypothesis 3*, we test supervisor behaviors that support or fail to support competence building. Competence building is defined as including management behaviors which either support or deter employees from education, training and other knowledge development activities. The behaviors in this category range from "refuses to commit resources and employee time for training and education activities" to "spends time discussing and implementing a learning plan with each employee".

Communication[20]

Bowen and Lawler (1992), Kanter (1983), Kimberley and Evanisko (1981), and Spreitzer (1995) discuss the importance of non-hierarchical, flexible structures in encouraging employee empowerment and creativity. We have included these organizational dimensions in *Hypothesis 4*, where we test employee perceptions of their supervisor's communication and decision-making behaviors. Here we have included behaviors that support or fail to support open, candid and non-hierarchical approaches to sharing of ideas between the supervisor and the employee, the employee with other managers, and the employee with different divisions in the organization. The supervisory behaviors range from "reinforces organizational hierarchies by insisting that employees be of the same level (does not want employees to talk to other managers)" to "listens openly and attentively to suggested improvements in how s/he does his/her job and often adopts the suggestions".

Information Dissemination

Information is described as one of the foundations for innovation in the psychological empowerment literature as well as being highlighted as important in the development of trust in the relationship between the employee and supervisor (Bowen and Lawler, 1992; Spreitzer, 1995). In *Hypothesis 5*, we test employee perceptions of supervisor behaviors which support information dissemination, including the open sharing of information both about company policies and goals, and difficult subjects, like layoffs and restructurings. Behaviors in this area range from "our group is often the last to know about changes in the company because our manager does not tell us things" to "clearly explains the reason for the organization's goals or policies and forewarns employees about expected changes whenever possible".

Note that we have differentiated the categories of communication and information dissemination by the following. Included in the communication category are the behaviors and attitudes which reflect the openness of the supervisor to input from subordinates and open flow of communication from subordinates to others in the organization. Included in the information dissemination category are behaviors and tools which the supervisor uses to share general or environmental information concerning corporate goal, events, and activities with subordinates.   Thus, in our definition, communication includes issues of authoritarian and hierarchical approaches, whereas information dissemination is concerned with issues of providing information that could enhance the employees' participation in the organizational priorities of the firm.

Rewards and Recognition

Bowen and Lawler (1992) and Lawler (1973; 1990) indicate that rewards and recognition can positively influence employee involvement in organizations.   *Hypothesis 6* tests the influence of supervisory use of rewards and recognition on employee environmental initiatives. Here we included behaviors that reinforced employee participation using formal awards, monetary rewards and informal recognition of employee activities. The behaviors in this area range from "I have seen my manager publicly reprimand another employee" to "uses bonus pay or other monetary awards to reward employees who have achieved or surpassed their goals". Note that rewards and recognition are closely linked to organizational and individual goals in the literature, but we have separated management of goals and responsibilities into a separate category as the management behaviors differ significantly in the two categories.

Goals and Responsibilities

Goal clarity can contribute to creativity (Abbey and Dickson, 1983; Andrews, 1975; Ellison, James, McDonald, Fox and Taylor, 1968; Pelz, 1956; Witt and Beorkrem, 1989). Supervisors can "facilitate creativity by enhancing subordinate motivation through the application of mechanisms such as goal setting" (Redmond, Mumford, and Teach, 1993; p. 122). Since the manager's role in setting measurable and clear goals can positively affect employee creativity, *Hypothesis 7* tests the influence of goal setting and sharing of responsibility for organizational goals in encouragement of employee initiatives. The behaviors in this area of the survey tool range from "tries to manage every detail of an employee's work so that the employee has little freedom to do his/her job independently" to "involves employees in changes by instilling ownership of problems and responsibilities for solutions in every employee".

Supervisory Support for General Management versus Environmental Management

We have noted that the survey consists of behaviorially-anchored rating scales (BARS) of general management behaviors which support or fail to support subordinate initiatives. We asked two questions using the same six sets of behaviors in the categories above to determine if supervisory support in general has as much impact on employee environmental initiatives as supervisory support applied to environmental management. Thus, we have formed *Hypothesis 8*, which tests the influence of general management support versus environmentally specific support to see which has a more significant impact on employee environmental initiatives. We believe, from our review of the literature and observations in companies that support specifically for environmental activities will have a more significant and larger impact, but we test this empirically in our study.[21]

2.4   SUMMARY OF CONCEPTUAL FRAMEWORK CHAPTER

Our conceptual framework tests for employee perceptions of organizational support for environmental initiatives (*Hypothesis 1*) using thirteen environmental policies, and using six categories of supervisory behaviors, tests for employee perceptions of supervisory support for environmental initiatives (*Hypotheses 2-8*).

In this chapter, we described our process for selecting thirteen environmental policies and defined these. Then we defined the six categories of supervisory behaviors. Our conceptual framework presents the following eight hypotheses, which are tested by our empirical investigation:

Hypothesis 1: Organizational support, in the form of a well-communicated environmental policy, will be positively related to employees' willingness to promote ecoinitiatives.

Hypothesis 2: Supervisory behaviors that support employees' innovation will be positively related to their willingness to promote ecoinitiatives.

Hypothesis 3: Supervisory behaviors that support employees' competence building will be positively related to their willingness to promote ecoinitiatives.

Hypothesis 4: Supervisory behaviors that support employees' communication with others will be positively related to their willingness to promote ecoinitiatives.

Hypothesis 5: Supervisory behaviors that support dissemination of information to employees will be positively related to their willingness to promote ecoinitiatives.

Hypothesis 6: Supervisory behaviors that support employees using rewards and recognition will be positively related to their willingness to promote ecoinitiatives.

Hypothesis 7: Supervisory behaviors that support employees by managing goals and responsibilities will be positively related to their willingness to promote ecoinitiatives.

Hypothesis 8: General management support behaviors from supervisors will have less affect on employee eco-innovation than supervisory support behaviors in the area of environmental management.

Chapter 3

Research Methodology

In this chapter we describe the research methodology used in this investigation. First, the design of our study is presented. The data sources are defined and the sample selection criteria given. The process by which the behaviorally-anchored rating scales (BARS) were developed is described. Then we present the independent and dependent survey measures. Finally, the analytical models used for the data analysis are discussed.

3.1 STUDY DESIGN

The purpose of this study was to obtain an understanding of which organizational and supervisory support factors significantly increase the likelihood that employees would have tried to promote environmental initiatives in their companies. Eight hypotheses were offered concerning factors which may or may not have a significant impact of employee willingness to eco-innovate. To test these eight hypotheses an empirical survey of employees was conducted. The data from this survey was analyzed using logit estimation, likelihood ratio test and Chi-square test of differences techniques, as well as descriptive statistics.

Data Sources and Sample Procedures

The data for this study were collected from employees of European companies, selected because of their proactive environmental policies. This created a voluntary selection bias, which we discuss below. The sample used for construct validation was composed of 353 mid and low-level employees. These employees worked at six companies head-quartered in Europe, each of which was listed by national ranking organizations within the top two hundred companies in terms of sales. The companies, employing 1500[1] to 41,000 people, came from a variety of industries, including chemical, entertainment, manufacturing, medical devices, oil, and retail. The employees were randomly selected from diverse work force units representing different functions, divisions and geographic locations.[2] Employees were based in twelve countries: Austria, Belgium, Canada, Finland, France, Germany, Italy, The Netherlands, Norway, Spain, United Kingdom, and United States. The survey instrument was available in English, German, Dutch and Finnish. (Note that 6.8% (24 of 353) of the sample respondents were from the United States or Canada as one company had operations in North America.) We were not interested in cross company comparisons, but rather wanted to test our hypotheses by obtaining a diverse sample of employees working in different countries.

The environmental pressures and issues facing the companies in the sample were very different as they were head-quartered in a variety of countries and operating in different industries. We wanted a sample that would represent a diverse cross section of European companies, in order to be able to generalize our conclusions to European employees in work places with different pressures and approaches. The companies had in common a strong commitment to environmental sustainability policies. (Below we discuss the process for selecting these companies.) As a result, finding statistically significant results will be all the more important since our sample will have no specific industry or country effect, allowing us to argue that it is representative of more than one country and industry.

The respondents were assured of confidentiality.[3] Surveys were returned directly to the authors for processing, and only aggregate results were reported back to the organizations. Data were collected in 1996 and 1997. 1465 surveys were distributed. The response rate was 24 percent.

We chose companies which environmental strategy journals noted as having proactive environmental management programs and that had published environmental policies.[4] We interviewed environmental managers to confirm that each company had a strong commitment to its environmental program. By this careful screening of companies, we ensured that the employees in the sample worked for organizations which were more supportive of environmental activities than the average European head-quartered company. Note that there is a selection bias in the sample of companies as we approached twelve companies who met our criteria and six of these were willing to participate.

In addition to the positive (pro-environmental protection) bias of the companies in the sample, we also believe there was a positive self-selection bias in the sample of employees. We assumed that employees responding to the survey were likely to be interested in and/or informed about environmental issues than the population of employees from which they were drawn.[5]

While we assume the sample is positively biased, this bias toward support for self-described environmental initiatives was intentional. In order to examine the relationships between environmental policies, supervisory support behaviors and employee initiatives, we selected a sample of organizations where such policies and behaviors were likely to exist. In this way we could test employee perceptions of signals from the organization and their supervisors. Note that the positive biases in the sample would tend to reduce variance in responses. Notice that lower variance may give us less statistical significance. Therefore finding such significance using this positively biased sample will be all the stronger.

Thus, our sample is representative of environmentally-oriented employees from large companies with head-quarters in Europe. These companies had international operations and took a proactive approach to managing the environmental impacts of their businesses.

At the same time we conducted the survey with employees, we conducted a separate survey of environmental managers at the same six companies. We asked if their company had the environmental policies listed in questions 1 through 13 of the survey questionnaire. We did this in order to compare the employee responses with those of the environmental management professionals of the company. This allowed us to find out if employees were knowledgeable about which policies actually existed. All six companies had policies for questions 1-7 and 13; one company did not have a policy for 8 (use of Life Cycle Analysis), 9 (management understands sustainable development and is addressing it), and 11 (toxic chemical use reduction); three companies did not have a policy for 12 (reducing use of unsustainable products); and, four companies did not have a policy for 10 (fossil fuel use reduction). This environmental manager policy survey allowed us to better interpret the results for hypothesis 1.

Survey Development Process

The survey was developed over a one year period beginning in January 1996 using a process which included a literature review, focus groups and other consultations with company managers and employees. During this process a set of thirteen environmental policy factors and a unique set of behaviorally-anchored rating scales were developed. The validity of the questionnaire tool was tested.

Part One of the Survey: Environmental Policy Questions

The first part of the survey questionnaire consisted of a list of thirteen environmental policies. The development process for this list was discussed in Chapter 2. Please refer to Table 2.2 for the complete list of environmental policies.

Part Two of the Survey: Behavioral Questions

The second part of the survey questionnaire was developed using the literature on work performance measurement regarding behaviorally-anchored rating scales (BARS) (Farr, Enscore, Dubin, Cleveland, and Kozlowski, 1980; Hampton, Summer, and Webber, 1987; Landy and Farr, 1983; Smith and Kendall, 1963). As the literature indicates, this type of measurement tool makes for a more precise assessment process by filtering the value judgment into an empirical set. And it helps evaluators refrain from offering speculations about causes of behaviors. Instead of asking the employee to provide a general opinion about the supervisor, the employee is asked to select, from a universal set of behaviors, which they believe is most typical of their supervisor. We have used the BARS instrument in a novel way. While there is a long history of using BARS for traditional manager assessment of employee work performance, we instead use the tool for an employee assessment of their manager's work behaviors that support environmental initiatives.

Using the process described in Farr, Enscore, Dubin, Cleveland, and Kozlowski (1980) we developed the six BARS from behaviors in the categories of Innovation, Competence Building, Communication, Information Dissemination, Rewards & Recognition, and Goals & Responsibilities.

Step 1 in Development of BARS: Literature Review As described in Farr, Enscore, Dubin, Cleveland, and Kozlowski (1980), we began the development of BARS for supervisor behaviors with a review of the literature. This literature review indicated organizational characteristics which supported employee empowerment and creativity, provided work environment support for innovation, and relied on management support behaviors to encourage employees. (Please refer to literature in Chapter 1, such as: Amabile, Conti, Coon, Lazenby, and Herron, 1996; Argyris and Schon, 1978; Campbell and Cairns, 1994; Cummings, 1965; Delbecq and Mills, 1985; Garvin, 1993; Pearn, Roderick, and Mulrooney, 1995; Peters 1990-91; Redmond, Mumford, and Teach, 1993; Senge, 1990; Spreitzer, 1995; Wagner, 1991.) We were interested in discovering if those management behaviors that supported employee creativity and innovation in general, would be the same behaviors that supported employee environmental innovation. We developed a list of general behaviors based on the literature review.

Step 2 in Development of BARS: Employee Interviews As proposed by Farr et al. (1980), the next step in the BARS validation process was interviews. We used information received at interviews with 50 employees at five companies to compare these general behaviors found in the literature with those that actually exist in companies that support employee environmental initiatives. The five companies were chosen from those noted in the environmental literature as supporting employee environmental involvement.

During 50, one-hour interviews we asked a series of questions in the categories to management structure, communication, information, learning, innovation, rewards and recognition, measurement of goals. These categories and questions were developed based upon the literature review. Thirty-six questions were asked with probing follow up questions used where appropriate to get further information on daily behaviors observed by the respondent from his or her supervisor. (Refer to Appendix 1.1 for Interviewing Questionnaire used during these sessions.) These interviews with employees confirmed that those organizational behaviors noted in the literature as supporting employees in general were also the behaviors that employees perceived to be supportive of their environmental initiatives.

Step 3 in Development of BARS: Allocating Exercise From these interviews, we developed a list of 132 behaviors which employees said provided effective and ineffective support for their environmental activities. We asked a group of 20 environmental managers to perform an allocating exercise with these behaviors. The managers allocated the 132 behaviors to the six categories of attributes of an organization that supports innovative actions by employees: Innovation, Competence Building, Communication, Information Dissemination, Rewards & Recognition, and Measurable Goals and Responsibilities. (Refer to Appendix 1.2 for the complete Allocating Exercise Questionnaire.) The group was provided with a list of Performance Characteristics of managers in Learning Organization. This list was derived from the literature to help them identify which category the behavior might fit into. (See page 2 of Appendix 1.2 for Performance Characteristics list.)

There were two purposes to this exercise. First, the group reduced the number of behaviors from 132 to 86. Following the process for construct validation suggested by Farr, Enscore, Dubin, Cleveland, and Kozlowski (1980), we used a sixty percent rule for this reduction. Farr et al. noted in their research paper that "behavioral examples which could not be allocated to a dimension using this rule were considered to be ambiguous and were eliminated from further phases" (p. 9). Therefore, only in cases where there was sixty percent agreement by respondents that the particular behavior belonged to a single attribute did we keep the behavior for the next step in the survey development process.

Second, the exercise provided the best match of behaviors to attributes. This pairing is the basis for the scaling exercise described in step four. For example, the behavior "gives incomplete or inaccurate information to employees" was consistently allocated to "Information Dissemination" so it was retained as one of the behaviors for the scaling exercise.

Step 4 in Development of BARS: Scaling Exercise   The outcome of the allocating exercise was converted into a document where each of the 86 behaviors was listed under one of the six categories (refer to Appendix 1.3 for this document). This document was used in a group of company employees in order to create ranked lists of behaviors for the final survey questionnaire. Employees assigned a value from 1-5 to each of the behaviors using 1 for highly unsupportive/ineffective managerial behaviors and 5 for highly supportive/highly effective behaviors. Employees at another company also performed the same ranking exercise. Together twenty employees from the two companies completed the scaling document.

The behavioral portion of the survey questionnaire was created out of the ranked/scaled lists. Specifically, we created ranked lists of behaviors under the six categories of Innovation, Competence Building, Communication, Information Dissemination, Rewards and Recognition, and Goals and Responsibilities from this scaling exercise. For example, under Innovation we retained ten behaviors with the ranked list beginning with the least supportive behavior of "Usually objects to changes and new ideas and finds excuses why they can not be implemented" followed by progressively more supportive behaviors. The last behavior on the ranked list of Innovation behaviors was "encourages partnerships with other departments in order to implement new ideas". Thus we see that the previously unranked list was made into a list ranked from least to most supportive supervisory behaviors using the responses from the scaling exercise. (Refer to Table 3.1 for a complete presentation of the six ranked lists that became the BARS we used in the final questionnaire.)

Step 5 in Development of BARS: Testing of Draft Questionnaire   This draft behavioral survey created from step 4's Scaling Exercise, coupled with the list of thirteen environmental policy questions, was then tested with employees at a large water utility company. Note that when respondents were answering the BARS questions, we asked that they select only one (most typical) behavior from each of the six lists of behaviors. The test involved 50 randomly-selected employees from six departments in the company, 24 of who accurately completed the survey questionnaires. (Response rate for the survey validation test was 48 percent.) Based on this test and feedback from a focus group with 15 environmental managers, the survey tool was revised.

For example, the focus group wanted the questionnaire to include four additional screening questions. First they suggested including two questions concerning the respondents general opinion of management and future prospects in the company. And they recommended including two questions concerning the financial health of the company: i.e. whether the company had had major lay offs in the last three years, and if the company was profitable and/or growing over the last three years. The focus group also asked for slight modifications on the wording of the instructions included in the questionnaire. (Refer to Appendix 1.4 to see the Final Survey Questionnaire.)

The final questionnaire was made available to six companies who conducted a survey of their employees. The survey questionnaires were returned directly to the author in postage pre-paid envelopes in order to assure respondents that no one in the company would see the questionnaires.

Table 3.1   Behaviorally-Anchored Rating Scales for Supervisory Behaviors

Ranked from least supportive to most supportive.

Innovation

My manager...

1. Usually objects to changes and to new ideas and finds excuses why they can not be implemented.
2. Seldom experiments with new ideas or methods of doing things.
3. I would never approach my manager with a suggested change because I know s/he would be angry with me for interfering in his/her area of responsibility.
4. Neither encourages nor discourages new ideas from employees.
5. Gives feedback to employees on their ideas and suggestions, even if they are not adopted.
6. Would accompany an employee to discuss and promote the employee's idea to another manager.
7. Experiments with new ideas in order to examine whether they are profitable/feasible to adopt on a large scale.
8. When someone makes a mistake, we usually discuss, as a group, how to avoid the problem in the future.
9. Sends employees to other locations in the company and elsewhere to learn about innovative processes and other ways of doing business.
10. Encourages partnerships with other departments in order to implement new ideas.

Competence Building

My manager...

1. Refuses to commit resources and employee time for training and education activities.
2. Delays giving employees the training/education they need when they change their job functions.
3. Neither encourages nor discourages employee participation in training and education.
4. Usually encourages participation in any appropriate learning situation in which an employee would like to engage.
5. If there is something new I need to know, my manager will make sure I have training/education on it quickly.
6. Realigns employee responsibilities to allow employee time for training, site visits, or exploring new techniques for doing his/her job.
7. Spends time discussing and implementing a learning plan with each employee.

Table 3.1 continued

Communication

My manager...

1. Reinforces organizational hierarchies by insisting that employees be of the same level or the level immediately above in order to communicate. (i.e. Does not want employees to talk to other managers.)

2. Avoids difficult discussions and ignores problems as they are developing.

3. Listens to employees, then forgets/ignores what they have said.

4. Neither encourages nor discourages employee communication.

5. Encourages employees to express concerns about company decisions and policies so that the concerns can be openly discussed.

6. Creates an open environment in which to discuss decisions that affect the business. For example, welcomes employee discussions about possible changes, improvements or problems that need solving.

7. Answers questions honestly even if the answer is not what the employee wants to hear.

8. Listens to and values input from employees and managers from all parts of the company.

9. Listens openly and attentively to suggested improvements in how s/he does his/her job and often adopts the suggestions.

Information Dissemination

My manager...

1. Our group is often the last to know about changes in the company because our manager does not tell us things.

2. Tries to hide unpopular decisions and information from employees.

3. Gives incomplete or inaccurate information to employees.

4. Keeps information about problems in our area private from the rest of the company and tries to solve them without help.

5. Neither actively aids nor hinders information flow to employees.

6. Uses information systems, such electronic bulletin boards, videos, computer systems, etc. to share information amongst employees.

7. If there is too much information, s/he sets priorities and establishes what is most important for employees to know.

8. Encourages employee trust by openly announcing information without delay about troubling situations, like lay-offs or restructurings.

9. Clearly explains the reason for the organization's goals or policies and forewarns employees about expected changes whenever possible.

Table 3.1   continued

Rewards and Recognition

My manager...

1. I have seen my manager publicly reprimand another employee (or my manager has criticized me in front of others).

2. If I do a good job, I am not certain that my manager will notice. But, if I make a mistake, I am sure s/he will notice and probably criticize me for the mistake.

3. Seldom, if ever, rewards or recognizes an employee for work well done.

4. Neither recognizes nor discourages employee contributions.

5. If the company does well, my manager will reward all of his/her employees.

6. Looks for opportunities to praise positive employee performance, both privately and in front of others.

7. Rewards a good idea by implementing it and giving the responsible employee(s) credit.

8. Uses company award systems to recognize particularly good performance of employees.

9. Uses bonus pay or other monetary awards to reward employees who have achieved or surpassed their goals.

Management of Goals and Responsibilities

My manager...

1. Tries to manage every detail of an employee's work so that the employee has little freedom to do his/her job independently.

2. Seldom talks to employees about goals and responsibilities, except when required by company policies.

3. Is often vague about what s/he wants from an employee.

4. Neither encourages or discourages employees from taking responsibility.

5. Keeps responsibility for all decisions.

6. Delegates specific tasks to employees and tells them precisely how the tasks should be performed.

7. Talks regularly with employees to assess progress toward explicit employee goals.

8. Tells an employee right away when there is something wrong with his/her work.

9. Uses both quantitative (numbers) and qualitative (quality) measures to assure individual is making progress toward or contributing to company goals.

10. Involves employees in changes by instilling ownership of problems and responsibilities for solutions in every employee.

3.2   SURVEY MEASURES

Screening Measures

Four screening questions were included in the questionnaire at the request of the managers attending the final focus group. They wanted these questions asked in order to determine if any of the respondents were negatively biased going into the survey. Sources of bias against management or against the company were measured.

The first two screening questions attempted to identify if the respondent had a negative bias toward management:

- What is your general opinion of management in your company?

- How do you feel about your future prospects in the company?

Respondents were given a choice of answers: "very poor", "poor", "average", "good" and "excellent".

Two discrete choice screening questions were asked in the questionnaire to determine if the respondent felt that the company was in poor economic health, possibly resulting in poor morale on the part of the employee:

- Has the company had major layoffs during the last three years?

- Has the company been profitable and/or growing over the last three years?

Respondents had a choice of "yes" or "no".

Note that we found these screening measures to yield no significant results, a fact we discuss in more detail in our chapter on data analysis.

Environmental Policy Measures

Thirteen environmental policies were used to measure employee perceptions of organizational support for environmental activities in the firm. There were a range of policies statements provided in the questionnaire beginning with a question regarding the employee's knowledge of the existence of a published environmental policy and ending with policies of sustainable development and social equity. The first policy question asked about employee knowledge of the general published environmental policy, whereas, each of the other twelve policies were more specific, sub-policies of this main environmental policy statement. We used this design in order to test if specific policies, like environmental purchasing, training, fossil fuel use reduction, etc. had a significant influence on employee environmental initiatives. (Refer to Table 2.2 for a complete list.)

We were interested in measuring the impact of employee knowledge of the existence of these policies, and their perceptions of organizational commitment to these policies. In our survey, each of the thirteen policies was listed as an affirmative statement. From the responses we tested for 1) whether or not the employee was aware that the policy exists (knowledge of the existence of the environmental policy), and 2) employee perception of company commitment to the said policy. These two measures used a five-point scale (Strongly Agree, Partially Agree, Don't Know, Partially Disagree, and Strongly Disagree). Employee knowledge of existence of the policy was measured by the answer of "agree" (meaning that the employee knows that this policy exists), "don't know" (meaning that the employee doesn't know if this policy exists) or "disagree" (meaning that the employee does not think that this policy exists). Employee perception of organizational commitment was measured with the descriptors "highly or partially" agree or disagree. (For instance, we measure whether the employee "highly agrees" that the policy exists, if there is an increased probability that the employee had tried an environmental initiative compared to other responses, like "partially agrees", etc.)

Respondents to the survey were asked to select one of the following choices of five discriminant answers for each of the thirteen policy questions: "strongly agree" that my company has such a policy (5); "partially agree" that my company has such a policy (4); "don't know" if my company has such a policy (3); "partially disagree" that my company has such a policy (2); or "strongly disagree" that my company has such a policy (1).

These measures were used as in order to provide us with more information than a "yes" or "no" discrete choice response. From the a response of 4 or 5, we assume, that "yes" the employee believes the policy exists. But, from a 5 we also assume that the employee believes that the management is strongly committed to the policy, whereas a response of 4 tells us that the employee, while s/he believes the policy exists, s/he only partially agrees that the company is committed to such a policy. The opposite reasoning holds true for the response of 1 or 2. Both responses mean that the employee does not believe the policy exists. But a response of 1 means that the employee strongly believes that the company is not committed to any such policy, whereas the weaker 2 response means that the employee only partially believes that there is a lack of commitment to the policy. A response of 3 is a neutral response, meaning that the employee does not know if such a policy exists and does not have strong positive or negative feelings about the company's commitment to such a policy.

BARS Measures

Six BARS were developed using the five step process described above. These measures were rank ordered lists of behaviors falling into six categories: Innovation, Competence Building, Communication, Information Dissemination, Rewards and Recognition, and Management of Goals and Responsibilities. There were ten behaviors listed for Innovation; seven for Competence Building; nine for Communication; nine for Information Dissemination; nine for Rewards and Recognition; and ten for Management of Goals and Responsibilities.[6] Each of the lists were ranked from least supportive to most supportive behaviors. Employees were asked to select a single behavior from each list which they felt was most typical of their direct supervisors behaviors. A single behavior was selected which was most typical of the supervisor when managing general work responsibilities of employees. Then, from the same list, a single behavior was selected which the employee felt was most typical of their supervisor when managing employee environmental-related responsibilities.

From the results of the employee survey, we tested seven hypotheses for these BARS. Using the logit model for binomial regressions, the first six hypotheses (2-7) were regressed to see whether or not supervisory support behaviors in each of these categories had a statistically significant impact on employee willingness to promote an environmental initiative. We also performed a likelihood ratio test to see if the group of six BARS had a statistically significant impact on the dependent variable. Then for each of the six BARS we did a third test (using a Chi-Square test of differences) for *Hypothesis 8*. This test compared the responses in each of the six areas to identify if there was a significant difference between employee selection of behaviors for general management versus environmental management. Then the logit model was used to identify if there was a significant difference in the impact on employee eco-initiatives.

Number and Fill-in-the-Blank Measures

Another measure was taken in the survey. Respondents were also asked to assign a discrete number from 1-5 (1 = unsupportive, 5= highly supportive) which described their direct supervisors' level of support. These numbers were compared to the behavioral choice for the same category. This numerical information provided a baseline with which to compare behavioral responses. (Note: we did not find this number to be a helpful measure for interpreting the results. We discuss this finding in Chapter 4.)

Finally, the employee was invited to describe specific examples of supportive or unsupportive behaviors they had observed when the supervisor was managing environmental activities of employees. This final section of fill-in-the-blank responses was reported confidentially to the companies who participated in the survey. This information gave companies an opportunity to learn about their areas of strengths and weaknesses.[7]

Dependent Variable Measure

We used a discrete choice dependent variable in our survey. We asked respondents if they had ever "tried to promote an environmental initiative within the company". They were given a choice of a "yes" or "no" response to this question. Responses to this question were used as the dependent variable in our econometric analysis of the statistical results.

In our study, we were interested in self-reported eco-initiatives. We believe that these give an accurate reflection of the reality within the company of the willingness of personnel to promote environmental innovation. In other words, if an employee reported that s/he had tried to promote an environmental initiative, it is very likely that s/he actually did as reported. Thus, this single measure is a very good proxy for actual employee eco-innovation in companies.

An environmental initiative was broadly defined. It was seen as any action taken by the employee that he or she thought would improve the environmental performance of the company by either decreasing the environmental impact of the company, solving an environmental problem in the company, or developing an environmental product/service for the company.[8]

Eco-initiatives can vary significantly in time, effort, scope and importance, but this is not of concern to us, as we were interested in studying the relationship between employee perceptions of organizational and supervisory support and their willingness to apply creative solutions to the environmental area. Thus, the only two issues of importance are the ones we measured, 1) whether or not the employee attempted to implement an eco-initiative, and 2) whether perceived support, either from organizational policy or supervisory behaviors, had an impact on the willingness of the employee to eco-innovate. Thus, we believe that our dependent variable was sufficiently complex for the issue we were interested in measuring.

Please note that our survey was designed for the average company employee, many of who were working in blue-collar jobs on the shop floor. And, the survey was conducted in four different languages with employees from twelve different countries. Therefore, our definitions and language used in the questionnaire needed to be understandable to people of different levels of education, sophistication, and from different cultural backgrounds. This explains the relative simplicity of our definitions and the nature of the descriptive language used in our research.

We understand that often dependent variables are independently verified (using a separate survey instrument, or interviews), but in our study we were interested in whether or not the respondent felt he or she had tried an environmental initiative, not whether someone else knew that they had. We were testing the employees' perceptions of their own actions and their perceptions of their supervisors behaviors that supported or failed to support the employee's environmental creativity and innovation. Therefore, we believed that the best approach was to ask the respondents the question directly.

Therefore, we did not independently verify the employee responses because of the difficulty in doing so, but because there is likely to be a strong correlation between an employee's self-perception that s/he has tried an environmental initiative and someone else believing that that employee had done so. And, what is instructive for our study is the fact that the employee believed that s/he had tried the initiative, not the fact that someone else perceived that they had.

Measurement of Interactions between Two Sets of Independent Variables

We are also concerned about the possibility of interaction between our two sets of independent variables, policies and BARS. We examined this issue with the whole range of analytical techniques used to examine the impact of each set of variables separately.

We present the statistical models used for our analysis.

3.3   ANALYTICAL MODELS

The three main models, beyond classical descriptive statistics, used for the statistical analysis of our data were the logit model, likelihood ratio test, and Chi-square test of differences. We present each below, describing our choice of these models and the formulas used in the analysis of the survey data.

The Logit Model

We needed to select a model which could be used to analyze data with a discrete dependent variable. "Discrete" means that the dependent variable is non-continuous. In our case, the dependent variable was the answer of "yes, I have tried to promote an environmental initiative in my company" or "no, I have not tried to promote an environmental initiative in my company". Discrete dependent variables do not lend themselves to the use of linear regression analysis, such as Ordinary Least Squares (OLS) or Generalized Least Squares (GLS).

A linear regression would impose a relationship between a dependent variable y and a set of independent variables $X = (x_1, x_2, \_., x_n)$ such that:

$$y = a + bX + e$$

where a is a constant and b is the set of coefficients relative to the set of independent variables. e is an error term which is assumed to be normally distributed. So that the estimated values of y are:

$$y = a + bX$$

But using this type of linear regression model would have had several drawbacks. The first is that it would produce "non-sense probabilities" or "negative variances" (such as negative probabilities, for example) (Greene, 1993; p. 635). The second draw back from using the linear regression technique with our data is the problem described by Cramer (1991) of the need to assign "complex properties" to the random e term, thus taking away the simplicity, which is the main appeal of the linear regression model (p. 6).

Because probabilities are bounded by 0 and 1, in order to analyze the probability that an employee will try to promote an environmental initiative, it did not make sense for us to use the classical linear model. Instead we used a qualitative response model which links the decision (i.e. yes or no) to a set of factors (policies and BARS). In our case, we needed to use a binomial model.

There are two standard models used for this type of discrete choice analysis, the probit and the logit models. The main difference between these two models is that the logit model uses a logistic distribution density, whereas probit uses a Normal distribution density. Both densities are symmetrical. Cramer (1991) explains: "Logit and probit functions which have been fitted to the same data are virtually indistinguishable, and it is impossible to choose between the two on empirical grounds" (p. 17). And since "probit is less tractable analytically than logit" (p. 15), Cramer (1991) suggests the use of logit instead. Thus, we decided to use the logit binomial discrete choice model as the main model for our data analysis. Note that the logit model is also the most used discrete choice model in academia and has well behaved statistical properties, which allow for easy convergence of maximum likelihood estimation of the model.

The logit model allows us to study the impact of a set of independent variables on y, the dependent variable which takes a value 1 if an employee had tried an initiative and 0 otherwise.

The logit model imposes a relationship between a set of independent variables x (environmental policies or supervisory behaviors) and a dependent variable y (whether or not the employee had tried an environmental initiative) that only takes values 0 or 1, such that:

$$\hat{y} = \Pr(y = 1 | X) = \frac{1}{1 + e^{-(\alpha + \beta X)}}$$

Likelihood Ratio Test

We perform likelihood ratio tests in order to test the null hypothesis of no incremental significance of using chosen sets of independent variables versus only using a constant or using another restricted set of variables. In other words, likelihood ratio tests allow us to test for the significance of multinomial logit analysis even in situations where, because of multicolinearity, interpretation of the significance of the coefficient of each independent variable is not possible.

The test statistic is

$$LR = 2(\log L(\theta_1) - \log L(\theta_2))$$

with L(θ) the likelihood function corresponding to the set of parameters of the set of variables indicated by the indices (2 being by necessity nested in 1). This statistic is asymptotically distributed as a chi square with r degrees of freedom, with r equal to the difference in the number of variables considered in the two sets (the number of restrictions).

The Chi-Square Test

We used the Chi-square test to see whether or not there was a fundamental difference between supervisory behaviors when managing general business issues versus when managing environmental issues, with respect to the answers given in the survey. We set up the test supposing that for a given question we have a given number of employees **x** for each alternative answer **j**, and this, for each company **i**: For example for Innovation BARS:

Answers for Innovation BARS:	1^a	2	3	–	10^b	Sum per line
General Management	$x_{1,1}$	$x_{1,2}$			$x_{1,10}$	
Environmental Management	$x_{2,1}$	–				
Sum per column						Total sum

[a] least supportive behavior
[b] most supportive behavior

First, we calculated the sums for each of the six characteristic BARS (innovation, competence building, communication, etc.) for each answer:

$$A_j = \sum_i x_{ij}$$

Second, we computed the sum for each of the characteristics BARS for all answers:

$$B_i = \sum_j x_{ij}$$

Summing the sums for each of the six characteristics gave us the total amount of observations:

$$T = \sum_i B_i$$

Then, we calculated the theoretical frequencies as:

$$F_{ij} = \frac{B_i}{T} A_j$$

Finally, we compute the following expression:

$$D = \sum_i \sum_j \frac{(x_{ij} - F_{ij})^2}{F_{ij}}$$

which follows a Chi-square distribution with degrees of freedom:

df = (number of answers-1) i.e. 9 for the example proposed here.

On this basis, the test informs us on whether there is a significant difference between employee perceptions of behaviors their supervisors use to support general versus environmental management.

3.4   SUMMARY OF METHODOLOGY CHAPTER

In this chapter we examined the methodological tools used to study our research question. We presented a description of criteria used to select our sample, including which types of employees and companies we surveyed. Here we discussed our sampling procedures and possible sources of bias within the data. We then gave a step by step description of our survey development process highlighting the thorough use of both theory and consultations with managers and employees to create and validate the BARS portion of the survey instrument. We described our survey measures and gave our reasons for selecting these types of measures. Finally, we presented our choice of analytical models and formulas for the analysis of the data, including a discussion of their relevance for use in our research.

Chapter 4 presents the analysis of the data collected during our survey.

Chapter 4

Analysis of Survey Results

In this chapter we present the results of the empirical investigation. First, we discuss the objective of the analysis. Second, we present statistical analyses of the survey results. Then, we discuss conclusions we have drawn from our analysis of the results.

4.1 SUMMARY OF THE OBJECTIVE OF THE ANALYSES

The objective of the analysis was to examine the relationship[1] between employee environmental initiatives and environmental policies (*Hypothesis 1*), and the relationship between employee environmental initiatives and supervisory behaviors (*Hypotheses 2-8*). We tested these eight hypotheses using logit analyses (Cramer, 1991; Greene, 1993), Chi-square test of differences, likelihood ratio tests, and descriptive statistics, like mean, standard deviations, and correlation tables (Cooper and Emory, 1991).

In all cases the dependent variable was whether or not the employee had tried to promote an environmental initiative.

Our general approach was to use logit analyses to determine the nature of the impact of our independent variables on our dependent variable.

The independent variables for hypothesis 1 were thirteen environmental policies. (Refer to Table 2.2.) And for hypotheses 2-8, the independent variables were the six BARS. (Refer to Table 3.1.) Employees were asked to use the same BARS to evaluate both the general and environmental behaviors of their direct supervisor. Note we had the employees evaluate both the general supervisory behaviors and those applied to the environmental management case in order to test if managers had significantly different behaviors when managing the daily business than when managing environmental activities. We were also interested in identifying those general management behaviors that had a significant influence on employee's willingness to promote an environmental initiative. We used a Chi-square test of differences (Cooper and Emory, 1991) to test if there was a statistically significant difference between supervisors' support behaviors in the general management versus environmental management cases.

In order to test for significant correlations between policies, between BARS and between the two sets of independent variables which could lead to multicolinearity problems, we performed correlation analyses.

We used likelihood ratio tests to test the null hypothesis of no incremental significance of using chosen sets of independent variables versus only using a constant or using another set of independent variables in our logit formula (Cramer, 1991). We also performed over 500 logit regressions to determine the nature of the interaction between the two sets of independent variables (policies and behaviors).

Note that we defined four screening measures that we included in the employee survey in order to control for biased general opinion of some employees. An analysis of these found no significant results; therefore we have not included a discussion of these measures below. Specifically, the first two screening questions were asked to determine if any of the employees in the survey had a correlation between negative opinion of general management in the company and poor future prospects in the company with their answers to the BARS questions. The second two screening questions were asked to determine if any employees in the survey felt that the company was in poor economic health, thus leading them to have poor morale, and resulting in a significantly different set of answers to the BARS questions than the other employees. We found that fewer than three percent of the employees answered "poor" or "very poor" to the two screening questions on "general opinion of management" and "future prospects in the company". Only six out of 353 questionnaires answered positively to the question on "major layoffs during the last three years in the company and negatively to whether the company had been "growing/profitable in the last three years". We performed a test of significance on the set of answers to the BARS questions from these employees, comparing them to the answers from employees who did not appear to have a negative opinion of management, future prospects in the company, or the economic health of the company. We found no statistically significant difference in responses to the BARS questions between the two groups. Therefore we decided that these employees, even if they did have negative feelings about their management or their company, did not create a bias in the data, and we included them in the final data analysis.[2]

Also note that we included in the survey an additional measure that we do not discuss below. Respondents were asked to assign a number between 1 and 5 to characterize how supportive their supervisors was in each of the six BARS categories. These numbers were much less useful than the behavioral questions. The main reason was that the numbers were less descriptive than behaviors and that almost all of the employees chose either a 3 or a 4 to describe their supervisor regardless of the behavior chosen. And, since there was no way of knowing the criteria used by each employee when assigning a number (i.e. what is the difference in criteria used for selecting a 3 versus a 4?) we chose to exclude this information from our data analysis.

4.2   RESULTS OF EMPLOYEE SURVEY: PART ONE

Hypothesis 1: Employee Perceptions of Organizational Support

Hypothesis 1: Organizational support in the form of a well-communicated environmental policy will positively influence individuals to try to promote eco-initiatives.

Here we test whether employees who perceive a strong organizational signal from a well-communicated policy of environmental protection are more likely to have tried to promote an environmental initiative. Specifically, we examine the impact of employee perceptions of each of thirteen environmental policies upon their willingness to promote environmental initiatives.[3]

Analysis of Descriptive Statistics for Environmental Policy Questions

Table 4.1 gives means, standard deviations, and correlations of the environmental policy questions. The mean and standard deviation for responses to the first environmental policy question show that respondents generally reported that they "strongly agreed" that their organization had a published environmental policy (question 1: mean 1.61; s.d. 0.73). The majority of the respondents also "strongly agreed" that their organizations had environmental performance targets (question 2; mean 1.46; s.d. 0.71), published an environmental report (question 3: mean 1.22; s.d. 1.01) and used an environmental management system (question 4: mean 1.22; s.d. 0.91). All of the companies in the survey had these four policies and the employee responses show that most respondents knew that such policies existed with the majority "strongly agreeing", showing that they felt there was company commitment to such policies.

Table 4.1 Descriptive Statistics and Correlation Table for Environmental Policies

Independent Variables	Mean	s.d.	1	2	3	4	5	6	7	8	9	10	11	12
1 Published environmental policy	1.61	0.73												
2 Specific targets for environmental performance	1.46	0.71	0.48***											
3 Publishes annual environmental report	1.08	1.01	0.41***	0.47***										
4 Uses environmental management system	1.22	0.91	0.37***	0.51***	0.45***									
5 Environmental considerations in purchasing decisions	0.75	1.05	0.36***	0.45***	0.33***	0.38***								
6 Employee environmental training	0.58	1.22	0.25***	0.36***	0.35***	0.32***	0.35***							
7 Employees responsible for company environmental performance	0.86	1.20	0.25***	0.34***	0.33***	0.37***	0.37***	0.53***						
8 Life cycle analysis	0.21	0.99	0.18***	0.23***	0.29***	0.21***	0.37***	0.28***	0.31***					
9 Management understands/addresses issue of sustainable development	0.66	1.01	0.32***	0.30***	0.35***	0.34***	0.39***	0.31***	0.41***	0.39***				
10 Systematically reduces fossil fuel use	0.14	1.27	0.08	0.23***	0.14**	0.31***	0.27***	0.15***	0.28***	0.25***	0.30***			
11 Systematically reduces toxic chemicals use	0.73	1.15	0.17**	0.29***	0.18**	0.22***	0.31***	0.05	0.13**	0.21***	0.35***	0.34***		
12 Systematically reduces consumption of unsustainable products	0.40	1.12	0.19***	0.23***	0.21***	0.20***	0.40***	0.16***	0.25***	0.37***	0.36***	0.51***	0.59***	
13 Applies same environmental standards at home and abroad	0.64	1.09	0.19***	0.24**	0.17***	0.19***	0.26***	0.31***	0.25***	0.18***	0.27***	0.03	0.14**	0.13*

Scale: Strongly Agree(2), Partially Agree(1), Partially Disagree(-1), Strongly Disagree(-2), Don't Know(0)

*** $p \leq 0.001$

** $p \leq 0.01$

* $p \leq 0.05$

There was less agreement of respondents on the remaining environmental policies. Even for policy questions 5, 6 and 7, where all companies had such policies, the majority of respondents "partially agreed" or "didn't know" (means: 0.75, 0.58, and 0.86; s.d.: 1.05, 1.22, and 1.20, consecutively). For questions 8, 9, 10, 11, and 12, where not all companies had such policies, we found lower means (ranging from 0.14 to 0.73, and larger s.d.'s (ranging from 0.99 to 1.27)). And for question 13 (on same environmental standards at home and abroad), a policy which all companies had, we found the mean (0.64) and s.d. (1.09) to be in the same range. This result shows that employees in the sample were less certain of the organization's commitment to/existence of these policies. Since policy questions 8-13 were those policies in the survey that specifically asked about the sustainable development policies of the company (use of LCAs, management understanding of sustainable development, reduced use of toxic chemicals, fossil fuels, unsustainable products, and same environmental policies at home and abroad), we ascertain that employees do not perceive a strong signal from the organization to work toward sustainable activities, except in the cases where the policy of sustainability is communicated in the overarching written environmental policy, the environmental targets, the environmental report and the environmental management system.

In general, the result of the descriptive statistical analysis shows that employees are less knowledgeable and tend to agree less strongly that the company has the specific policies of questions 5 through 13, than they are concerning policies 1, 2, 3, and 4. This conclusion is indicated by higher standard deviations and lower means for answers to such specific policies questions as 5 through 13. Yet, respondent knowledge of and belief in organization commitment to a written environmental policy is clearly indicated. And, though means and standard deviations show a weaker response for questions 2, 3, and 4 compared to the question on having a written environmental policy, employees also perceive a strong organizational signal from company environmental targets, reporting and management systems.

The correlation results show significant positive correlations between nearly all policies. All but three of the policies have positive correlations to each other at p-value of 0.05 or lower. We are not surprised by this result as we would anticipate that employee knowledge of different environmental policies might be correlated to one another as the method of communication of the general environmental policy and its different sub policies may be similar.

Note that these correlations lead us to perform two kinds of logit analyses. First we did an analysis looking at all of the coefficients simultaneously but found no significant results at below the p-value of 0.05. This result may come from the multicollinearity of the data as appears from the correlation tables. Interpretation is therefore difficult. Then we performed the logit taking one question at a time and found a number of statistically significant results, which we describe below.

Logit Analysis of Environmental Policy Questions

Tables 4.2 shows the results of logit analyses examining the relationships between the dependent variable (employee environmental initiatives) and the independent variables, employee perceptions of the thirteen environmental policies. We performed two separate regressions, one with and one without a constant. The regressions were performed taking each of the thirteen independent variables one at a time in order to avoid problems of multicolinearity, as mentioned above. Thus, we performed a series of bivariate regressions. The results shown in Table 4.2 (with a constant) explain the data better than our analysis performed without a constant. The reason that analyzing the data with a constant gives a better explanation of the results is that individuals in the survey can be assumed to have an initial positive (non-zero) probability to undertake an initiative regardless of their answers to questions 1 through 13.

The result of the logit analyses showed that there was a relationship (p-value = 0.02) between employees promoting an environmental initiative and their knowledge that a published environmental policy (question 1) existed. This result does not only show strong significance, but represents a strong impact too (as indicated by the strength of the coefficient at 0.37). This is a clear result that indicates that employees responding to the survey were more likely to have tried an environmental initiative if they strongly agreed that the company had a written environmental policy.

Looking at Table 4.2 we can see that there is a large difference in the size of the coefficients in quite a few cases, with the largest and most positive coefficient for policy 1 (0.37; written environmental policy) as compared to negative coefficients for policy 10 (-0.18; reduced fossil fuel use), policy 11 (-0.17; reduced toxic chemical use), and policy 11 (-0.16; reduced consumption of unsustainable products). We provide the following explanation for these differences. As noted above, some of the companies in the survey did not have policies for 8 (0.08; Life Cycle Analysis), 9 (0.09; management understands and addresses issue of sustainable development), 10, 11, and 12. Therefore, we would expect that

Table 4.2 Logit Analyses of Dependent Variable on Environmental Policy Independent Variables (with a Constant)

Constant (p-value)		Independent Variable	Coefficient (p-value)
-0.74 (0.01)*	1	Published environmental policy	0.37 (0.02)*
-0.31 (0.21)	2	Specific targets for environmental performance	0.13 (0.40)
-0.33 (0.04)*	3	Publishes annual environmental report	0.19 (0.08)
-0.13 (0.46)	4	Uses environmental management system	0.00 (1.00)
-0.14 (0.29)	5	Environmental considerations in purchasing decisions	0.02 (0.81)
-0.20 (0.10)	6	Employee environmental training	0.13 (0.15)
-0.14 (0.30)	7	Employees responsible for company environmental performance	0.02 (0.78)
-0.14 (0.20)	8	Life cycle analysis	0.08 (0.47)
-0.19 (0.14)	9	Management understands/addresses issue of sustainable development	0.09 (0.38)
-0.15 (0.16)	10	Systematically reduces fossil fuel use	-0.18 (0.04)*
0.00 (1.00)	11	Systematically reduces toxic chemicals use	-0.17 (0.07)
-0.07 (0.54)	12	Systematically reduces consumption of unsustainable products	-0.16 (0.11)
-0.23 (0.07)	13	Applies same environmental standards at home and abroad	0.15 (0.12)

* p-value ≤ 0.05   Dependent Variable = Employee Environmental Initiatives

there would be less impact from these factors on employee initiatives. The lower coefficients show this to be the case. Four of these policies do not have a statistically significant impact on employee eco-initiatives. The fossil fuels policy has a significant negative impact (p-value = 0.04), which is explained below.

Other policies with non-significant and relatively lower coefficients include policy 4 (0.00; uses an environmental management system) and policy 5 (0.02; environmental considerations in purchasing decisions). Since all companies have these two policies we interpret the low coefficients and non-significant impact to mean that having an environmental management system or an environmental purchasing policy does not directly influence employee willingness to eco-innovate. Larger coefficients and stronger p-values for policy 3 (0.19, p-value=0.08; publishes annual environmental report) and policy 6 (0.13, p-value=0.15; employee environmental training), on the other hand, indicates that these two policies have a larger chance of influencing employee environmental initiatives. But, as we have noted above, none of these policies has statistical significance below p-value 0.05.

We also wanted to analyze whether taken together the thirteen policies had a statistically significant impact on the dependent variable. Therefore, we performed a likelihood ratio test of significance considering the 13 policies together and found that the thirteen policies have a significant impact on eco-innovation at p-value = 0.00 even though individual coefficients cannot be interpreted because of multicolinearity.

Test of Extreme Values to Determine Strength of Significant Results

Table 4.3 compares the probabilities that employees would have tried to promote an environmental initiative if they strongly agreed that a policy existed versus strongly disagreed that such a policy existed. In fact, the probability, calculated from the logit formula, that respondents who strongly agree that the company has a published environmental policy will try to promote an eco-initiative is 50% versus only 19% for those who strongly disagreed. (To see this result, refer to Table 4.3, which shows a test of extreme values.) The result of the comparison for question 1 shows that having a well communicated environmental policy which employees believe the company is committed to is the single most important policy factor that influences employee eco-initiatives. This is an important result, as from it we confirm our hypothesis that one of the drivers of employees' eco-initiatives is most certainly the published environmental statement, in

Table 4.3 Test of Extreme Values: Environmental Policy Questions
(Comparing Strongly Disagree (-2) to Strongly Agree (2) Responses)

Environmental Policy Questions	Strongly Disagree	Probability Employee Tried Environmental Initiative	Strongly Agree	Probability Employee Tried Environmental Initiative
1 Published environmental policy.	-1.47	18.69%	0.00	49.98%
2 Specific targets for environmental performance.	-0.57	36.19%	-0.06	48.57%
3 Publishes annual environmental report.	-0.71	33.07%	0.04	51.10%
4 Uses environmental management system.	-0.13	46.70%	-0.13	46.73%
5 Environmental considerations in purchasing decisions.	-0.19	45.32%	-0.09	47.78%
6 Employee environmental training.	-0.45	38.88%	0.05	51.34%
7 Employees responsible for company environmental performance.	-0.19	45.38%	-0.09	47.85%
8 Life cycle analysis.	-0.30	42.53%	0.02	50.42%
9 Management understands/addresses issue of sustainable development.	-0.38	40.68%	0.00	50.00%
10 Systematically reduces fossil fuel use.	0.21	55.15%	-0.51	37.51%
11 Systematically reduces toxic chemicals use.	0.34	58.53%	-0.34	41.48%
12 Systematically reduces consumption of unsustainable products.	0.24	56.02%	-0.38	40.55%
13 Applies same environmental standards at home and abroad.	-0.54	36.81%	0.08	51.94%

the absence of which, employees are far less likely to put their creative energy toward environmental improvements in the workplace.

In addition, Table 4.3 shows that in ten out of thirteen of the policy questions, respondents have a higher probability of having tried an environmental initiative if they strongly agreed that the company had such a policy. The three policies where this is not the case are questions 10 (fossil fuel use reduction), 11 (toxic chemical use reduction), and 12 (reduced use of unsustainable products). We interpret this result to mean, first, that positive knowledge of /belief that the company is committed to the environmental policies usually has a positive relationship to a respondent's willingness to try to promote an environmental initiative. And, secondly, that companies do not have policies of 10, 11, and 12 in many cases (one company did not have a policy for 11 (toxic chemical use reduction); three companies did not have a policy for 12 (reducing use of unsustainable products); and, four companies did not have a policy for 10 (fossil fuel use reduction)), resulting in a weaker relationship between the dependent and independent variables.

Overall, we find that the increased probability that an employee had tried an environmental initiative if they answered "strongly agree" to the policy question indicates that having such policies, and showing commitment to them can positively affect employee willingness to eco-innovate. Note that none of these results (except written environmental policy question 1 and the fossil fuel reduction question 10) are statistically significant. But as we will see below, in the section analyzing interactions between the two sets of independent variables, these other policies do have an indirect effect on the dependent variable. (Where employees perceive a strong signal from the policies, they are more likely to perceive supervisory support for eco-initiatives and to have tried to promote an environmental initiative.)

Explanation of Other Significant Result: Fossil Fuels Use Reduction

Looking at Table 4.2 again, we can see that the more specific sub-policies in questions 2-13 are perceived by employees to be less important than the company having shown commitment to a general written environmental policy (question 1). This can be seen by the weaker coefficients and weaker relationships to the dependent variable (p-value $> .05$). There is one exception. In the case of a policy of fossil fuel use reduction which was negatively related to employee environmental initiatives, we can see a stronger relationship to the dependent variable (p-value $= 0.04$) In four of the six companies in the survey, the environmental managers independently verified that no such "fossil fuel reduction" policy existed in their companies. Thus, it is possible employees who were well informed about the lack of a company fossil fuel policy also tended to be those who were more likely to try an environmental initiative (hence the negative coefficient).

This result could indicate that those employees who are better informed about the organizational signal of environmental policy (whether the signal is positive or negative) may also be those who are more likely to use that information to encourage their environmental initiatives. This supposition is reinforced by a separate logit analysis we performed with question 1 (general policy) and question 10 (fossil fuel policy). (Note that there was no significant correlation between the two questions in Table 4.1.) This analysis showed that there was a statistically significant relationship even when both independent variables were included in the logit formula (p-values: 0.01 for Q1; 0.02 for Q10. Coefficients: 0.41 for Q1; -0.20 for Q10). Question 1 still has a significantly higher impact on employee eco-innovations as its coefficient is twice as large as that of question 10. This result leads us to believe that while fossil fuel use reduction is important to sustainable development as described in the literature (Cramer, 1996a; Robert, 1989), it has not yet been adopted by even the set of proactive companies selected for our survey. Since only two companies have a plan to systematically reduce the use of fossil fuels, we place less significance on the result for question 10 than on question 1, where all six companies had such a policy. Nonetheless, this result indicates that even in a time where it is acknowledged by companies that fossil fuels are a limited, non-renewable natural resource, the environmentally concerned companies in our survey have not yet taken a proactive approach to reducing fossil fuel use.

Conclusion from Testing Hypothesis 1: Relationship between Environmental Initiatives and Environmental Policy

In conclusion, we found that our hypothesis that environmental policy can provide a signal to employees, which can positively affect their willingness to promote an environmental initiative, is proven in the case of having a written environmental policy. Those employees that did not strongly agree that the company had a general environmental policy were far less likely to try to promote an eco-initiative. Those employees that did believe strongly that such a written policy existed were 50 percent more likely to have tried an initiative than those who strongly disagreed that the policy existed. The logit analysis indicates that effective communication of the environmental policy positively affects employee eco-innovation. And, employee belief in company commitment (organizational commitment) to environmental protection as shown by a well communicated policy can positively affect eco-innovation.

Second, we found that, though a weaker, non-statistically significant result than in the case of question 1, respondents who "strongly agreed" that their company had policies 2 through 9 and 13 were more likely to have tried an environmental initiative than those who "strongly disagreed". Note that in these cases, though not statistically significant, the coefficient was always positive. This indicates that a belief in organizational commitment to these policies can positively influence employee willingness to try to promote eco-initiatives. But, as noted, these factors do not have a statistically significant influence on the dependent variable, which indicates that this result should not be over interpreted.

Third, questions 10, 11 and 12 on policies to reduce use of toxic chemicals, fossil fuels, and unsustainable products did not have a positive influence employee eco-initiatives. Our explanation for this result is two-fold. First, some companies in the survey did not have policies for questions 10, 11, and 12. Second, companies, possibly as a result of this ambivalence toward these policies, were less likely to have communicated these policies to their employees (see the low mean and high standard deviations for questions 10, 11, and 12 in Table 4.1). Thus, while the literature indicates the importance of these policies for corporate sustainability, there is little evidence that the policies exist or act as a positive signal for employees to eco-innovate.

In conclusion, having a written, well communicated environmental policy has a strong and significant direct impact on the dependent variable, but none of the other eleven environmental policy factors do (excluding fossil fuels reduction). Therefore, we believe, based on this result, that organizational encouragement for employee environmental initiatives is best communicated through a single overarching environmental statement, which is broadly disseminated to employees. (Later in this chapter when we look at the interaction between the two sets of variables we will see that the other environmental policies do have an indirect effect on the dependent variable.)

4.3   RESULTS OF EMPLOYEE SURVEY: PART TWO

Hypotheses 2-8: Employee Perceptions of Supervisory Support

Hypotheses 2-7: Supervisory encouragement, as demonstrated by daily behaviors that support employees' environmental activities, will positively influence individuals to eco-innovate. These support behaviors are the same as generally encourage employee innovation in general, and are in the areas of innovation (2), competence building (3), communication (4), information dissemination (5), use of rewards and recognition (6), and management of goals and responsibilities (7).

Hypothesis 8: General supervisory encouragement will not have as large an impact on employee environmental initiatives as supervisory encouragement in the area of environmental management.

In the second part of the survey we test whether employees who perceive supportive behaviors from their supervisors are more likely to try to promote an environmental initiative in their company.

Analysis of Descriptive Statistics for Supervisory Support BARS Questions

Table 4.4 gives the mean, standard deviations and correlations for the behaviorally-anchored rating scales (BARS) we created for supervisory behaviors. All means have been normalized to a scale of 10 even though we had different numbers of behaviors from which to chose for the six different BARS. We can see that the environmental means tend to be slightly lower than the general management means. This could indicate that direct supervisors being assessed in the survey were perceived by their employees to be less supportive of environmental activities than when managing general management activities. We test this possibility below using a Chi-square test of differences. Also note that the means tend to be above average (average = 5) indicating that most of the respondents perceived their supervisors to use more supportive than unsupportive behaviors in both the environmental and general management cases. And, we observe that the standard deviations are consistently large (ranging from 1.79 to 2.67) for all 12 questions. This leads us to conclude that there was a large variance in the responses, i.e. that some supervisors were perceived to use much more supportive behaviors than others being assessed in the survey.

We find that all of the independent variables have statistically significant positive correlations in the BARS (p-value$\leq$ 0.001). This is not

Table 4.4 Descriptive Statistics and Correlation Table for Supervisory Behaviors (BARS)

Independent Variables	Mean	s.d.	1	2	3	4	5	6	7	8	9	10	11
1 Environmental innovation	6.02	2.30											
2 Environmental competence building	5.70	1.79	0.31***										
3 Environmental communication	6.26	2.08	0.41***	0.41***									
4 Environmental information dissemination	6.95	2.13	0.33***	0.43***	0.51***								
5 Environmental rewards/recognition	5.93	2.25	0.30***	0.40***	0.47***	0.45***							
6 Environmental management goals/responsibilities	5.85	2.60	0.35***	0.38***	0.54***	0.50***	0.54***						
7 General innovation	6.49	2.21	0.51***	0.25***	0.38***	0.27***	0.27***	0.38***					
8 General competence building	5.84	1.79	0.22***	0.50***	0.32***	0.31***	0.37***	0.29***	0.23***				
9 General communication	6.28	2.18	0.24***	0.33***	0.58***	0.42***	0.42***	0.41***	0.34***	0.32***			
10 General information dissemination	7.09	2.31	0.26***	0.37***	0.40***	0.68***	0.37***	0.40***	0.27***	0.39***	0.48***		
11 General rewards/recognition	6.00	2.54	0.27***	0.26***	0.38***	0.36***	0.71***	0.38***	0.29***	0.35***	0.40***	0.37***	
12 General management goals/responsibilities	6.21	2.67	0.28***	0.29***	0.43***	0.43***	0.39***	0.65***	0.38***	0.34***	0.41***	0.47***	0.42***

Scale: Maximum = 10; Minimum = 1

*** p-value $\leq$ 0.001

a surprising result since management behaviors may be related to one another. For instance, a supportive manager would most probably have higher ratings on many of the variables, whereas a less supportive manager may have consistently lower ratings across the BARS. Due to our concerns about multicolinearity resulting from highly correlated variables, we performed two kinds of logit analyses. First we did an analysis looking at all of the coefficients simultaneously but found no significant results at below the p-value of 0.08. This result could clearly come from the multicollinearity of the data. Thus we performed the logit taking one question at a time and found a number of statistically significant results, which we describe below.

We also wanted to analyze whether taken together the twelve BARS had a statistically significant impact on the dependent variable. Therefore, we performed a likelihood ratio test of significance considering the BARS responses together and found that taken together, policies have a significant impact on eco-innovation at p-value = 0.00 even though individual coefficients cannot be interpreted because of multicolinearity.

Chi-Square Test Comparing Responses to Environmental and General Management Questions

By comparing the means in Table 4.4 we find that supervisors are relatively less supportive of employee environmental activities than they are of employees in general business activities. We used a Chi-square test of differences to verify whether this conclusion was valid (Cooper and Emory, 1991). (See Table 4.5 for results of Chi-square test.) The Chi-square test compared the environmental versus the general management supervisory behaviors in the areas of innovations, competence building, communication, information dissemination, rewards and recognition and management of goals and responsibilities. We got a statistically significant result at p-value $\leq$ 0.006 in all cases. The result of the test indicated that employees consistently selected less positive behaviors from the BARS when answering the environmental question, indicating that they felt their supervisors used less supportive behaviors than when managing general business issues. This was further verified by an analysis of the most frequent responses on the environmental behaviors questions. Consistently one of the most frequently selected response was that the supervisor "neither encourages, nor discourages" the employee's environmental activities in all six areas of innovation, competence building, communication, information dissemination, rewards and recognition, and management of goals and responsibilities.

Table 4.5 Chi-square Test of the Differences: Environmental versus General Management Behaviors

Environmental vs. General	Empirical Chi-square value	Theoretical Chi-square value	p-value
Innovation	23.02	16.92	0.006***
Competence Building	24.58	12.59	0.000***
Communication	27.17	15.51	0.001***
Information Dissemination	29.70	15.51	0.000***
Rewards/Recognition	40.43	15.51	0.000***
Management Goals/ Responsibilities	28.07	16.92	0.001***

*** p-value ≤ 0.001

Logit Analysis of BARS Questions

In Table 4.6 we show the results of multiple logit analyses examining the relationship between the dependent variable (employee environmental initiatives) and the independent variables of BARS for supervisory behaviors. We performed the regressions taking each of the twelve (six environmental and six general) independent variables one at a time in order to avoid problems of multicolinearity which result from the correlated variables, as mentioned above. We also made the regressions with a constant (see Table 4.6 for results). The results shown in Table 4.6 (with a constant) explain the data better than our analysis performed without a constant. The reason that analyzing the data with a constant gives a better explanation of the results is that individuals in the survey can be assumed to have an initial positive (non-zero) probability to undertake an initiative regardless of their answers to the BARS questions.

The results of these logit analyses showed that, for 5 of the 6 environmental questions, employees who perceived supportive behaviors from their supervisors showed an increased probability of trying to promote an environmental initiative (significant at p-values ≤ 0.05). Specifically, employees were more likely to have tried to promote an environmental initiative if they perceived supervisory support in the areas of environmental innovation, environmental competence building, environmental communication, use of environmental rewards and recognition, management of environmental goals and responsibilities. General innovation, general rewards and recognition, and management of general goals and responsibilities also had a positive relationship to employee environmental initiatives (significant at p-value ≤ 0.05).

Interestingly though these general categories of behaviors have systematically lower significance than the environmental categories of behaviors. We confirmed this by performing likelihood ratio tests on multinomial logit regressions. If taken together, both general BARS and environmental BARS variables had a strong impact on eco-innovation, as a likelihood ratio test shows (p-value = 0.00) (although individual coefficients could not be interpreted due to multicolinearity, as noted above). Adding general BARS to environmental BARS has no significant impact on eco-innovation (p-value = 0.82), as the likelihood ratio test shows, while adding environmental BARS to general BARS is highly significant (p-value = 0.00). We conclude that overall environmental BARS matter more and subsume the impact of general BARS on eco-innovation.

Table 4.6 Logit Analysis of Dependent Variable on Supervisory Behaviors Independent Variables

Constant (p-value)		Independent Variables	Coefficient (p-value)
-0.79 (0.01)**	1	Environmental innovation	0.11 (0.02)*
-0.98 (0.01)**	2	Environmental competence building	0.15 (0.01)**
-1.09 (0.00)**	3	Environmental communication	0.16 (0.00)**
-0.61 (0.10)	4	Environmental information dissemination	0.07 (0.17)
-0.93 (0.00)**	5	Environmental rewards/recognition	0.13 (0.01)**
-0.81 (0.00)**	6	Environmental management goals/responsibilities	0.12 (0.01)**
-0.77 (0.02)*	7	General innovation	0.10 (0.05)*
-0.42 (0.25)	8	General competence building	0.05 (0.39)
-0.67 (0.05)*	9	General communication	0.08 (0.09)
-0.70 (0.05)*	10	General information dissemination	0.08 (0.09)
-0.81 (0.00)**	11	General rewards/recognition	0.11 (0.01)**
-0.77 (0.01)**	12	General management goals/responsibilities	0.10 (0.01)**

*p-value ≤ 0.05   Dependent Variable = Employee Environmental
**p-value ≤ 0.01   Initiatives

Analyzing in more detail using binomial regressions, we also note that the coefficient of each of the significant general categories is lower than the coefficient of the corresponding environmental category, demonstrating lower impact. (Please refer to Table 4.6.) Several of the general behaviors are shown to be important to support employee eco-initiatives, but even more important, as the statistical analysis shows, are the environmental support behaviors.

The following argument is a possible explanation for this difference between the impact of general and environmental management support on employee willingness to eco-innovate. We can surmise that supervisors who used the more supportive behaviors in general, would also have shown support for employee activities in any reasonable business initiative in which the employee wished to participate. The impetus for environmental initiatives would not be coming from the supervisor in these cases, but rather from the employee, since the barriers to innovation and learning were removed freeing the employee to create solutions in the area of his or her choice. Thus, while the general support behaviors provided no direct encouragement for environmental initiatives, supervisors were sending a signal perceived by the employee to allow environmental, or other kinds of, solutions in the workplace. On the other hand, when the employee received a clear signal from the supervisor as to the desirability of eco-initiatives, employees would be more likely to apply their creative energy to finding solutions in the environmental area, reflecting not only their own intrinsic motivation, but also responding to organizational/supervisory extrinsic motivators. Thus, as we would expect, supervisory support in the area of environmental management had a greater significance and impact on employee eco-initiatives. There is support for this theory in Hostager, Neil, Decker, and Lorentz's (1998) work on employee intrapreneurial behaviors that lead to environmental innovations. (Please refer to Chapter 1 for further explanation of Hostager et al's theory of factors motivating eco-innovation.)

Specifically, we can see from these logit analyses, that while both general innovation (coefficient = 0.10: p-value = 0.05) and environmental innovation behaviors (coefficient = 0.11; p-value = 0.02) have a positive influence on employee willingness to eco-innovate; specific behaviors that support environmental innovation have more influence as can be seen by the stronger coefficient and p-value. And, while both use of general rewards and recognition (coefficient = 0.11; p-value = 0.01) and environmental rewards and recognition (coefficient = 0.13; p-value = 0.01) positively influence the dependent variable; specific behaviors that use environmental rewards and recognition have more influence as can be seen by the larger coefficient. And, finally, while both general management of goals and responsibilities (coefficient = 0.10; p-value = 0.01) and environmental management of goals and responsibilities (coefficient = 0.12; p-value = 0.01) have a significant positive influence on employee eco-innovation; management of environmental goals and responsibilities has a greater influence, as can be seen by the higher coefficient. We can see that while environmental competence building (coefficient = 0.15; p-value 0.01) and environmental communication (coefficient = 0.16; p-value 0.00) have a positive influence on employee willingness to eco-innovate, but neither general competence building or general communication have a significant influence. Finally, we see that supervisory support for environmental information dissemination does not have a significant influence on the dependent variable (coefficient = 0.07; p-value 0.17).

Thus, we find that the logit analyses verify the importance of supervisory behaviors in five of the six behavioral areas. Behaviors that support employee innovation in organizations as asserted by the literature and described by employees in excellent practice companies are behaviors that seem to have a significant influence over employee willingness to eco-innovate. The exception is the area of environmental information dissemination, which our logit analysis shows not to have a significant positive relationship.[4] And, we find that, while less important than environmentally-specific support behaviors, support behaviors in the areas of general innovation, rewards and recognition and management of goals and responsibilities also have a significant positive, but less strong, relationship to employee willingness to eco-innovate.

Test of Extreme Values to Determine Strength of Significant Results

Table 4.7 shows a test of extreme values, comparing the least supportive supervisory behavior in each of the six categories to the most supportive behavior. These calculations use the logit formula. Overall, we can observe an increased positive probability that employees will try an eco-initiative if they answered the most supportive behavior as compared to answering the least support behavior. This indicates that positive perception of management support in all twelve cases has a positive influence on employee activities. Said a different way, employees that perceive that their direct supervisor is highly supportive (uses the most supportive behaviors listed in the survey) are those employees who have the highest probability of having tried an environmental initiative. Without a doubt this points to the conclusion that the perception of supervisory support has a positive affect on eco-innovation in companies.

Table 4.7 Test of Extreme Values: Behavioral Responses
(Comparing Weakest Behavior vs. Strongest Behavior Responses)

Supervisor Behavior Questions *Supervisory Support for:*	Least Supportive Behavior (1)	Probability Employee Tried Environmental Initiative	Most Supportive Behavior (7, 8, 9, or 10)	Probability Employee Tried Environmental Initiative
Environmental innovation	-0.67	33.78%	0.34	58.45%
Environmental competence building	-0.83	30.42%	0.55	63.47%
Environmental communication	-0.94	28.15%	0.47	61.74%
Environmental information dissemination	-0.54	36.80%	0.10	52.49%
Environmental rewards and recognition	-0.80	31.01%	0.40	59.76%
Environmental goals and responsibilities	-0.69	33.41%	0.38	59.27%
General innovation	-0.67	33.79%	0.21	55.22%
General competence building	-0.37	40.88%	0.09	52.30%
General communication	-0.58	35.87%	0.18	54.50%
General information dissemination	-0.61	35.02%	0.11	52.67%
General rewards and recognition	-0.70	33.21%	0.30	57.26%
General goals and responsibilities	-0.67	33.84%	0.26	56.55%

Specifically, Table 4.7 shows that strong supervisory support of environmental innovation brings the probability of an employee's eco-initiative from a minimum level of 33.5% (with minimum support) to a level of 58.5% (with maximum perceived support). Even more impressively, strong environmental competence building support brings the probability of an employee's eco-initiative from a minimum level of 30.5% (with minimum support) to a level of 63.5% (with maximum perceived support). And strong support from the supervisor in the area of environmental communication brings the same probability from a minimum level of 28% to a maximum level of 62% (with maximum perceived support). In the area of environmental information dissemination, strong supervisory support brings the probability to 52.5% compared to 36.8% for minimal support. The trend is the same for environmental rewards and recognition (59.8% versus 31.0%) and management of environmental goals and responsibilities (59.3% versus 33.4%).

In comparison, general competence building has less impact with the probability of an employee's eco-initiative only reaching 52.3% with maximal general support versus 63.5% with maximal environmental support. Hence, general behavioral support, although it can often be helpful to generate eco-initiatives, is not as effective as supervisory behaviors directed specifically at environmental management. Thus, we conclude that companies who want employee eco-initiatives can do so most effectively by encouraging supervisory support for environmental activities of employees. Environmental communication has the strongest impact, followed by environmental competence building, but information dissemination does not seem to be directly effective.

Exception to Hypothesized Results: Environmental Information Dissemination

There is one exception in the behavioral results that should be noted. The area of environmental information dissemination (i.e. the sharing of specific environmental information to employees) had no apparent influence over employee willingness to promote environmental initiatives. While we had hypothesized that dissemination of information of environmental activities of the firm would positively impact employee willingness to promote an environmental initiative, we found that this was not the case.

There is some support for this negative result in the literature. Steger (1998a) has asserted that pressures from information coming from outside the organization may influence employee behaviors more strongly than internal information communication. Therefore, we provide the following possible explanation for this exception. External pressure to address environmental impacts of company activities may be perceived by employees, possibly shaping their environmental values and their interest in finding environmental solutions. We could see this as societal pressure for them to act. Whereas, on the other hand, internal information on environmental issues and company performance may not provide any direct signal to employees that the company would find it desirable for them to take environmental actions. Possibly, only when the environmental performance of the company is linked to individual/personal goals and responsibilities, coupled with the capability to act, would this information create an impetus for eco-initiatives by an employee. We surmise that this linkage between environmental information provided by the supervisor and individual environmental goals might be missing. Or, possibly, the response may indicate that supervisors may not be delivering environmental information at all.[5]

Conclusion from Testing Hypotheses 2-8: Relationship between Environmental Initiatives and Supervisory Support Behaviors

Even though we selected six environmentally proactive companies, we found that supervisors in the surveyed companies are perceived by employees to be less supportive of environmental activities than they are of general management activities. We also found that environmental support behaviors in five of the six categories (innovation, competence building, communication, use of rewards and recognition, and management of goals and responsibilities - but not information dissemination) positively affect employee willingness to try to promote an environmental initiative. The factor category of environmental information was not found to be significant. The results of our test of extreme values showed that employees who perceive highly supportive behaviors from their supervisors are more likely to have tried to promote an environmental initiative than those who perceived unsupportive behaviors from their supervisors. This result confirms our hypothesis that supervisory support behaviors are important to successful encouragement of employee eco-initiatives, notably environmental communication and environmental competence building. We conclude that supervisory behaviors tested for in the employee survey can have a positive influence on employee eco-innovation.

We found that while less important than environmental support behaviors, general management support behaviors from supervisors can have a positive relationship to employee eco-initiatives. Yet, hypothesis 8 was supported by the analytical results. Environmental management support behaviors by supervisors had a larger impact on employee eco-initiatives than general supervisory support behaviors.

4.4 INTERACTIONS BETWEEN TWO SETS OF INDEPENDENT VARIABLES

A separate set of tests, including correlation significance, likelihood ratio tests, and logit regressions, were performed on the two sets of independent variables (IVs) to determine if there were interaction effects between environmental policy and supervisory behavior variables, and if there were any, whether the effect was additive or substitutive.

Correlations Between Independent Variables

We have created a table that shows the correlations between the thirteen environmental policy variables and the twelve BARS variables. (Please refer to Table 4.8.) From this table we can see that the correlations between these two sets of independent variables are small in value, ranging from 0.00 to 0.31 with only one of the 171 correlations at 0.31 and the majority at 0.12 or below. Most of the p-values for these relationships were also low.

Of particular interest are the two environmental policies that had a statistically significant impact on the dependent variable. For both Policy 1 (written environmental policy) and Policy 10 (fossil fuel use reduction) the values are quite small, ranging from 0.00 to 0.13. Policy 1 had correlation p-values of 0.05 or below in four of the six cases with the other two p-values, for environmental competence building and rewards and recognition BARS at p-value = 0.16, and 0.07 respectively. Policy 10 had no p-values of significance at or below 0.05.

We interpret this result in the following way. Since all the values of the correlations are quite low we feel that there is very little correlation between the two sets of independent variables. This is a markedly different result than we see in Tables 4.1 and 4.4, which show that the thirteen environmental policies are highly correlated to one another and the BARS are highly correlated to one another. Thus, while we had a high probability of multicolinearity between policies and policies, and between BARS and BARS, we do not have the same concerns about multicolinearity between all policies and all BARS.

Table 4.8   Correlation Table for Environmental Policies and BARS of

Environmental

		Innovation	Competence Building	Communication	Information Dissemination
	1 Published environmental policy	0.122*	0.100	0.114**	0.126*
	2 Specific targets for improving	0.051	0.016	0.115*	0.123*
	3 Publication of environmental report	0.133**	0.137*	0.161**	0.155**
	4 Uses environmental management system	0.026	0.099	0.075	0.167**
	5 Environmental considerations in purchasing decisions	0.100	0.152**	0.094	0.173***
	6 Employees environmental training	0.201***	0.263***	0.231***	0.239***
Policies	7 Employees responsible for company environmental performance	0.197***	0.256***	0.187***	0.164**
	8 Life-cycle analysis	0.121	0.221***	0.109***	0.107*
	9 Management understands/addresses issue of sustainable development	0.273***	0.314***	0.266	0.256***
	10 Systematically reduces fossil fuel use	0.041	0.117	0.111	0.086
	11 Systematically reduces toxic chemicals use	0.087	0.115*	0.068	0.084
	12 Systematically reduces consumption of unsustainable products	0.111	0.155**	0.129***	0.134**
	13 Applies same environmental standards at home and abroad	0.204*	0.219***	0.249	0.250***

*** $p \leq 0.001$
** $p \leq 0.01$
* $p \leq 0.05$

Supervisory Behaviors

Variables

Rewards/Recognition	Management Goals/ Responsibilities	General					
		Innovation	Competence Building	Communication	Information Dissemination	Rewards/ Recognition	Management Goals/ Responsibilities
0.078	0.100*	0.074	0.115*	0.061	0.055	0.045	0.142**
0.040	0.073	-0.025	0.098*	0.080	0.045	-0.019	0.028
0.061	0.064	0.117*	0.075	0.070	0.038	0.033	0.102*
0.064	0.093*	0.037	0.081	0.033	0.013	-0.014	0.085*
0.117*	0.109*	0.046	0.180***	0.026	0.078	0.060	0.078
0.274***	0.216***	0.137**	0.167***	0.136*	0.175***	0.207***	0.123**
0.119*	0.174**	0.188**	0.112	0.103	0.108*	0.054	0.143**
0.053	0.148**	0.108	0.145**	0.011	0.031	0.037	0.068
0.211***	0.163**	0.167**	0.275***	0.158**	0.159**	0.152**	0.199***
0.001	0.076	-0.038	0.132*	-0.014	-0.007	-0.036	0.006
0.110	0.010	-0.054	0.192***	0.094	0.018	0.095	0.065
0.126*	0.116*	0.004	0.179***	0.083	0.007	0.093	0.112*
0.240***	0.202***	0.146*	0.192***	0.186***	0.231***	0.255***	0.185***

Likelihood Ratio Tests Between Independent Variables

In order to compare the differences between regressing a constant and all combinations of the sets of independent variables we ran a series of likelihood ratio tests. The sets of independent variables were as follows:

- Six environmental BARS,

- Six general BARS,

- Twelve BARS taken together,

- Thirteen environmental policies,

- Thirteen environmental policies and the six environmental BARS,

- Thirteen environmental policies and the six general BARS,

- Thirteen environmental policies and all twelve BARS taken together.

In all cases we found that we rejected the null hypothesis that there was no difference between regressing a constant versus regressing any set of independent variables. This means that in all seven of these likelihood ratio tests we found that the independent variables had more of an impact than regressing a constant in the logit regressions. Specifically, all of our independent variables, no matter if they are taken separately or together have a significant impact on our dependent variable (employee willingness to promote an environmental initiative).

Furthermore, we ran some additional Likelihood Ratio tests to refine these results. We first tried to determine if there was any difference between the explanatory power of the two sets of independent variables taken together (policies plus BARS) and either the policies alone or the BARS alone. We found that in both cases we rejected the null hypothesis that there was no difference (p-values = 0.00 and 0.02 respectively). From these results we determine that since there is a difference in the explanatory power when the BARS and policies are taken together versus when either the policies or BARS are taken alone, it is better to have both policies and BARS in order to influence employee eco-initiatives. Said another way, it increases the explanatory power of the independent variables on the dependent variable when all twelve BARS and all thirteen environmental policies are included. Thus the two sets of independent variables have an additive effect, although as we will see underneath it is a conditional rather than linear effect.

Logit Regression for Two Sets of Independent Variables

We were then concerned about understanding the structure of the interaction that may arise between the two sets of variables at the individual variable level. We first built logits regressing the same dependent variable on one supervisory behavior and one policy variable, in all possible combinations (156 regressions). (Please refer to the results in Table 4.9 with one example taking policy 1 with the twelve BARS.) Note that the results are similar to those in Tables 4.2 and 4.6 above, with the independent variables coefficients' values and significance being only marginally (and non significantly) lower. An economically weak and statistically insignificant substitution effect between policies and supervisory behaviors seems present.

Suspecting a non-linear and more complex interaction effect, we then ran regressions for each of the BARS on sub-samples (two regressions for each BARS, one with a sample of employees who perceive that the company has an environmental policy and a second sample of employees who do not perceive that such a policy exists). We did this for each environmental policy and each BARS (312 regressions). The results are interesting. (Please refer to Tables 4.10 and 4.11.)

Table 4.9 Logit Analysis of Dependent Variable on Supervisory Behaviors and Policy 1

Constant (p-value)		Variable	Coefficient (p-value)	Coefficient on Policy 1 (p-value)
-1.23 (0.00)**	1	Environmental innovation	0.10 (0.04)*	0.32 (0.05)*
-1.49 (0.00)**	2	Environmental competence building	0.15 (0.02)*	0.34 (0.04)*
-1.57 (0.00)**	3	Environmental communication	0.15 (0.01)**	0.33 (0.04)*
-1.04 (0.02)*	4	Environmental information dissemination	0.06 (0.27)	0.32 (0.05)*
-1.43 (0.00)**	5	Environmental rewards/recognition	0.13 (0.01)**	0.32 (0.05)*
-1.25 (0.00)**	6	Environmental management goals/responsibilities	0.11 (0.01)**	0.30 (0.07)
-1.32 (0.00)**	7	General innovation	0.09 (0.07)	0.37 (0.02)*
-0.93 (0.03)*	8	General competence building	0.04 (0.56)	0.37 (0.02)*
-1.23 (0.00)**	9	General communication	0.08 (0.12)	0.37 (0.02)*
-1.25 (0.00)**	10	General information dissemination	0.08 (0.11)	0.36 (0.02)*
-1.35 (0.00)**	11	General rewards/recognition	0.11 (0.02)*	0.34 (0.03)*
-1.24 (0.00)**	12	General management goals/responsibilities	0.09 (0.03)*	0.32 (0.04)*

**p≤0.01
*p≤0.05

Table 4.10 Logit Analysis of Dependent Variable on Supervisory Behaviors when Employees Perceived Strong Environmental Policy

Constant (p-value)		Independent Variables	Coefficient (p-value)
-0.71 (0.03)*	1	Environmental innovation	0.12 (0.02)*
-0.83 (0.03)*	2	Environmental competence building	0.15 (0.03)*
-0.89 (0.02)*	3	Environmental communication	0.14 (0.01)**
-0.40 (0.31)	4	Environmental information dissemination	0.06 (0.30)
-0.88 (0.01)**	5	Environmental rewards/recognition	0.14 (0.01)**
-0.67 (0.02)*	6	Environmental management goals/responsibilities	0.11 (0.01)**
-0.65 (0.07)	7	General innovation	0.10 (0.07)
-0.17 (0.66)	8	General competence building	0.03 (0.65)
-0.48 (0.17)	9	General communication	0.07 (0.16)
-0.42 (0.25)	10	General information dissemination	0.06 (0.24)
-0.72 (0.02)*	11	General rewards/recognition	0.11 (0.01)**
-0.60 (0.04)*	12	General management goals/responsibilities	0.09 (0.03)*

**p $\leq$ 0.01
*p $\leq$ 0.05

Table 4.11 Logit Analysis of Dependent Variable on Supervisory Behaviors when Employees Did Not Perceive Strong Environmental Policy

Constant (p-value)		Independent Variable	Coefficient (p-value)
-0.40 (0.73)	1	Environmental innovation	-0.18 (0.42)
-2.11 (0.14)	2	Environmental competence building	0.14 (0.56)
-2.61 (0.08)	3	Environmental communication	0.21 (0.36)
-1.60 (0.30)	4	Environmental information dissemination	0.04 (0.87)
-0.64 (0.57)	5	Environmental rewards/recognition	-0.13 (0.53)
-1.76 (0.08)	6	Environmental management goals/ responsibilities	0.08 (0.61)
-1.80 (0.16)	7	General innovation	0.07 (0.72)
-2.31 (0.14)	8	General competence building	0.17 (0.53)
-2.99 (0.10)	9	General communication	0.25 (0.34)
-6.83 (0.02)*	10	General information dissemination	0.72 (0.05)*
-1.32 (0.20)	11	General rewards/recognition	-0.01 (0.94)
-2.02 (0.07)	12	General management goals/ responsibilities	0.11 (0.52)

**p≤0.01
*p≤0.05

Consider the most significant policy, the published environmental policy. For employees that agree that there is a written environmental policy, results are very similar to Table 4.6 results, with similar coefficients and similar significance. For the sample of employees who do not perceive that there is a written environmental policy (a small sample of 33 out of a total of 353), all significance of supervisory behavior is lost at p-value ≤0.05. For this sub-sample of employees, supervisory behaviors that support environmental innovation, competence building, communication, rewards and recognition and management of environmental goals and responsibilities have no impact on their decision to be eco-innovative. On the other hand, employees who agreed that the company had a written environmental policy AND who perceived positive support behaviors from their supervisors in the five environmental BARS areas listed above, were more likely to have tried to promote an environmental initiative in their company. Notice also that our previous conclusion remains that supervisory support of environmental information dissemination does not have a significant impact on employee willingness to eco-innovate.

Also interesting is the fact that these asymmetric results hold for the eleven policies that were not deemed significant in Table 4.2. Indeed the same pattern is found with all policies except for policy 10 (Fossil Fuel Reduction). These results seem to point out that policies, even when they do not have a direct impact on eco-innovation, may have an indirect one by making employees that perceive these policies sensitive to supervisory support. Notice that we do not find any such systematic result when conditioning the impact of policies on the BARS. The results are not symmetric this way. Conditioning our logit tests on employee perceptions of supervisory behaviors does not give as systematic an impact to policies. There is no consistent indirect effect to behaviors while there is one to policies.

In conclusion, although there is no strict substitution effect between policies and supervisory behaviors' impact on eco-innovation, the interaction effect cannot be deemed to be simply additive either. Interaction is highly non-linear and appears to be conditional: Supervisory behaviors affect eco-innovation amongst employees that perceive strong policies to be in place while it has little or no effect on other employees.

Notably we can also conclude from these results that employees who do not perceive strong environmental policies (for twelve of the thirteen policies) are those who are less likely to perceive supervisory support, and who are less likely to try an eco-initiative. Therefore, perception of a strong signal of organizational support from environmental policies seems to have an indirect effect on eco-innovation.

4.5   DISCUSSION AND CONCLUSIONS FROM SURVEY RESULTS

We conclude from the statistic analysis of our sample that employees were consistently more likely to have tried an environmental initiative if they perceived company/organizational support for a written environmental policy AND if they perceived strong support behaviors in the environmental management area from their direct supervisors.   This confirms that some of the factors we framed in the conceptual model (Figure 2.1) do have a statistically significant impact on the dependent variable.

First, our results show a strong linkage between the existence of a published environmental policy and the willingness of employees to try environmental initiatives, indicating that employees respond positively with creative ideas in the environmental area, if they perceive a strong commitment to the environment from the organization.   In fact, having a convincing environmental policy triples the probability of employee eco-initiatives!

Second, the results of the behavioral portion of the survey showed a consistent and statistically significant result for five of the six categories, indicating the employees were definitely more likely to try environmental initiatives when they perceived supervisory support for those environmental activities.   Of those, environmental communication and environmental competence building had the most impact.   The logit analyses indicate that both environmental and general management behaviors can have a positive relationship to employees' willingness to promote environmental initiatives, but, environmental support had a greater impact and significance level.

And, the survey results confirm our prediction that support behaviors used to support other innovation are those that encourage environmental creativity and innovation.

More specifically, we summarize the analysis as finding the following interesting results:

First, using binomial logit analyses we found that five statistically significant independent variables had the greatest influence on employees eco-innovation: i.e. having a written environmental policy and five types of behaviors used by supervisors to support employees' environmental activities.   These binomial logit regressions showed that eleven of the thirteen policies had no significant impact on employee eco-innovation, and that behaviors in one of the environmental support behavior areas (Information Dissemination) had no significant impact.   From this result we conclude that not all environmental policies have equal impact on the dependent variable. The order of importance of the impact of the BARS is as follows (based upon size of coefficient):

1) Environmental Communication (coefficient=0.16).

2) Environmental Competence Building (coefficient=0.15).

3) Environmental Rewards and Recognition (coefficient=0.13).

4) Management of Environmental Goals and Responsibilities (coefficient=0.12).

5) Environmental Innovation (coefficient=0.11).

We note that environmental information dissemination does not have a direct impact on employee eco-innovation. This is an interesting result as we tested the existing assumptions in the environmental management literature that all of these factors would have an impact, yet our empirical test of the assumptions made in this literature finds a different result. This result suggests that existing theory may be incorrect. We define environmental information dissemination as information that is shared by the manager with the employee pertaining to company environmental activities. The lack of statistical significance of these behaviors shows that employees in our study perceived managerial environmental information dissemination as relatively insignificant in affecting their willingness to eco-innovate. Possible explanations for this result could be that either the manner in which the information is disseminated or the types of information currently given out by managers in the six organizations in the sample were not helpful in encouraging environmental actions. One possible further explanation, beyond the argument that the information was perceived as either poorly communicated or irrelevant, could be that receiving information is less important than encouraging communication, competence development, rewarding actions, sharing responsibility and welcoming ideas from employees. I.e. that managerial behaviors that encourage employees to passively receive information have less impact than behaviors that encourage actions by employees. We believe this may be the best explanation for this result since we find that environmental communication from employees to managers is the most significant factor in explaining the dependent variable. (See ranked list above.)

Items ranked higher on the list have relatively more impact on the dependent variable (employee eco-innovation) than items ranked at the bottom of the list. For example, we know that the explanatory power of the first item on the list (environmental communication) is greater than the last item on the list (environmental innovation) because the coefficient is fifty percent larger. Since the list reflects results from a single sample, we do not believe the ranked order of this list should be over interpreted. In other words, we posit that environmental communication and competence building behaviors have greater impact than environmental goals and responsibilities or environmental innovation behaviors as the coefficients are strongly larger.   But, we do not believe that environmental communication behaviors have definitively more explanatory power than environmental competence building, or that management of environmental goals and responsibilities has definitively more explanatory power than environmental innovation behaviors.

The environmental management literature shows support for our conclusion that environmental communication and environmental competence building behaviors are extremely important in explaining employee willingness to eco-innovate (Wehrmeyer, 1996). Environmental communication behaviors in our study were managerial behaviors that supported open and non-hierarchical communication from employees. We posit that the existence of types of behaviors that support employee communication within organizations may be precursors to other supportive behaviors from managers. For instance, we surmise that employee perceptions of managerial openness to communication can create a dialogue which facilitates positive employee perceptions of managerial behaviors that support competence building, use of rewards and recognition, etc.

Also, our study highlights the importance that behaviors, which support employee environmental competence building, have in explaining employee willingness to eco-innovate. The environmental management literature notes that managerial support for developing competencies and skills related to environmental activities can have a strong influence on employee environmental activities. From our study, we surmise that support for environmental competence building may be an important precursor to the other managerial support behaviors, as employee skills development provides employees with the ability to participate in environmental activities that improve the performance of the firm. I.e. having the necessary competence comes before the actions on the part of the employees. And managers that demonstrate support for environmental competence building send a signal that they want employees to devote energy and resources to environmental activities. Rewarding environmental activities follows the environmental activity, and sharing environmental goals and responsibilities would also seem to naturally follow after employees have the skills necessary to participate. Thus other managerial support behaviors that are ranked lower on the list, based on the explanatory power of the coefficient, we believe, can reasonably follow from behaviors that support open and non-hierarchical employee communication and the development of employees' environmental skills.

Second, using likelihood ratio tests we found that the thirteen environmental policies taken together had a statistically significant impact on the dependent variable. And the twelve sets of behaviors in the BARS (environmental and general) taken together also had a statistically significant impact on the dependent variable. Furthermore, the twelve BARS behaviors and the thirteen policies taken together also had a statistically significant impact, compared to taking only behaviors or only policies. We conclude from our likelihood ratio tests that the policies, the behaviors, and the policies plus behaviors have an additive effect. We interpret this result to mean that a company who wants employee eco-innovation would do well to adopt both environmental policies that signal organizational support for sustainable development AND to encourage supervisors (line managers) to use behaviors that are supportive of employee environmental actions. Note that we found that the environmental behaviors add explanatory power whereas general behaviors do not, indicating that direct supervisor support for environmental actions has more impact on employee eco-innovation than general support behaviors.

Third, our Chi-square test of differences showed that even in environmentally-proactive companies, such as participated in our study, supervisors are not as supportive of environmental actions as they are of employees in general. And, since our study shows that environmental behaviors have a greater impact, we conclude that there is an opportunity to increase eco-innovation if companies use more supportive behaviors to encourage employees to participate in environmental innovations.

Fourth, the results of several hundred multinomial logit regressions showed that taking both policies and behaviors at the same time does not have a statistically significant impact on the interpretation of the results. In other words, the binomial results hold. But, when we created two separate samples and ran more tests we found that those employee who perceive environmental policies are more likely to perceive supportive behaviors from managers. Whereas, employees who did not perceive company support for environmental policies were less likely to perceive support from managers. We conclude that the policy signal has an important indirect effect on the dependent variable. It increases the sensitivity of an employee to supervisory support. And, perceiving the policy signal thus indirectly affected the employee's willingness to eco-innovate.[6]

We do not want to over interpret this result as there could be other explanations for the linkage between the two sets of independent variables. We surmise that it is possible that employees who have less supportive managers may also perceive less of a policy signal because their managers have not communicated environmental policies to them. (One main route of environmental policy information would normally be through the direct supervisor.) Thus, these employees with less supportive managers are not perceiving either the policy or the supervisory support signal. But, note, we have run the regressions conditioning the results on employee perceptions of behaviors and not found a symmetrical effect. Said in another way; employees who perceived strong supervisory support were no more likely to perceive strong policy signals. Therefore, we believe that our first interpretation of the results is probably correct, that a strong policy signal (for twelve of the thirteen policies, excluding fossil fuels) increases the likelihood that employees will perceive supervisory support for eco-innovation.

As we have noted above, the fossil fuel policy is a special case in our results. Only two of the six companies had this policy, but employees who knew that it did not exist were statistically more likely to try eco-initiatives. We have interpreted this result to mean that those employees who were most knowledgeable about the actual company policies were also those who were most likely to try an environmental initiative. This interpretation adds strength to our conclusion above that communication of a strong policy signal is an important driver of eco-initiatives as knowledge of these policies seems to positively impact employee willingness to eco-innovate.

Finally, we have found that using binomial and multinomial logit regressions makes no difference to one of our most interesting results, i.e. that environmental information dissemination does not have a statistically significant impact on employees environmental innovation. That information dissemination should have an impact, to our knowledge, has never been disputed in the environmental management literature. But, Steger (1998a), as we have mentioned, does make the point that information from outside the company may have a greater impact than supervisory information dissemination. Nevertheless, as noted above, we believe that this result shows that managers are 1) not making the information they disseminate relevant to the individual employees, or 2) the information that is available within the company is too general to act as a driver of individual eco-initiatives. But, we believe the best explanation for this result may be that it indicates passive receipt of information from managers is less important than how employees perceive their actions to be received by their supervisors. The only alternative explanation we can find for this result is that our survey tool was flawed, but since we tested the BARS in advance of producing the final questionnaire, we feel this is unlikely.

Specifically, results for each hypothesis in our conceptual model are summarized as follows:

Hypothesis 1: Policy 1 (written environmental policy) had a strong and significant impact on employee willingness to promote environmental initiatives. Policy 10 (reduced use of fossil fuels) had a significant and negative impact on employee willingness to promote environmental initiatives. The other eleven policies (2-9 and 11-13) did not have a significant direct impact, but may indirectly effect the dependent variable through their interaction with the BARS.

Hypothesis 2: Supervisory behaviors, which supported environmental innovation, had a strong and significant impact on employee willingness to promote environmental initiatives. (Ranked fifth in importance of impact on the dependent variable.)

Hypothesis 3: Supervisory behaviors, which supported environmental competence building, had a strong and significant impact on employee willingness to promote environmental initiatives. (Ranked second in importance of impact on the dependent variable.)

Hypothesis 4: Supervisory behaviors, which supported environmental communication and non-hierarchical approaches to managing, had a strong and significant impact on employee willingness to promote environmental initiatives. (Ranked first in importance of impact on the dependent variable.)

Hypothesis 5: Supervisory behaviors, which supported environmental information dissemination, did **not** have a significant impact on employee willingness to promote environmental initiatives.

Hypothesis 6: Supervisory behaviors, which used environmental rewards, awards and informal recognition, had a strong and significant impact on employee willingness to promote environmental initiatives. (Ranked third in importance of impact on the dependent variable.)

Hypothesis 7: Supervisory behaviors, which shared environmental goals and responsibilities with employees, had a strong and significant impact on employee willingness to promote environmental initiatives. (Ranked fourth in importance of impact on the dependent variable.)

Hypothesis 8: Supervisory behaviors supporting environmental management had a more significant impact on employee environmental initiatives than general management support behaviors of supervisors, but managers are more supportive of general actions than environmental actions by employees.

In Chapter 5 we summarize our results and discuss implications for research and practice.

Chapter 5

Summary of Findings and Implications for Research and Practice

In this chapter we review our research findings, summarizing the major conclusions we have reached from our empirical study. Then we highlight implications for research and practice. Finally, we discuss where further research might be of interest.

5.1 SYNOPSIS

The purpose of this study has been to identify which organizational and supervisory support factors can positively influence employees to try to promote environmental initiatives in businesses. Based on the empirical work of Amabile, Conti, Coon, Lazenby, and Herron (1996), which showed that employee creativity in the workplace is encouraged by organizational and supervisory signals, we have developed our conceptual model (Figure 2.1). Built on findings from an extensive study of the organizational behavior literature in the areas of innovation, organizational behavior, and the environmental management literature, our model had two major parts. It used employee knowledge of/belief in management commitment to environmental policies as the signal of organizational commitment, testing which of thirteen environmental policies influenced employee eco-initiatives. And, for the supervisory support signal, it used six characteristics found in organizational behavior literature (innovation, competence building, communication, information dissemination, use of rewards and recognition, and management of goals and responsibilities), testing which of six sets of BARS influenced employee eco-initiatives. We analyzed a data sample composed of 353 mid and low level employees from European companies from a range of industries in order to empirically prove or disprove our eight hypotheses.

The data were analyzed using multiple logit analyses, Chi-square test of differences, likelihood ratio tests, and descriptive statistics procedures. The results confirmed that employee knowledge of their company's written environmental policy and their belief that their company is committed to such a policy was not only statistically significant, but strongly increased the probability that the employee will have tried to promote an environmental initiative in the work place. And, the results showed that supervisory behaviors that support environmental innovation, environmental competence building, environmental communication, use of environmental rewards and recognition and management of environmental goals and responsibilities, indeed have a significant positive influence on employee willingness to promote an environmental initiative. Eleven of the thirteen environmental policies and one of the behavioral BARS (environmental information dissemination) were shown to have no significant influence on employee eco-initiatives.

Employees who felt strong signals of organizational and supervisory encouragement were more likely to have developed and implemented creative ideas that positively affected the natural environment. Specifically, employees who perceived the existence of an environmental policy in their organization, and learning organization-type support behaviors from their supervisors, were more likely to have tried to promote eco-innovations. But interestingly, our results also indicate that supervisors, even in our positively-biased sample of European companies with strong environmental commitment, used less supportive behaviors when managing environmental as compared to general business activities. Furthermore, our statistical results show that environmental support behaviors from supervisors are a better driver for employee eco-initiatives than more general support not aimed at environmental management, although even these environmental behaviors have a differentiated impact. And, we found that environmental policies may have indirect, or conditional effects, which influence employee perceptions of supervisory support for eco-innovations. Below we interpret the major findings in our research.

Interpretation of General Findings

One main result of this research is the empirical evidence it provides that supervisory behaviors which are found in the general behavior literature, when applied to environmental management, can have a positive influence on employee participation in eco-innovation, but do not necessarily do so. While the organizational behavior literature indicates that supervisory support is important for innovation and employee participation in general, and the environmental management literature clearly points to a significant support role for managers in encouraging employee environmental actions, the role of supervisory support has not previously been tested for using a factors model of the type we developed and implemented. Our results show that support behaviors on the part of direct supervisors had a statistically significant positive influence on employee willingness to promote an environmental initiative.

A second important result of this research is the empirical evidence it gives that organizational support, as demonstrated by commitment to and communication of a written environmental policy, can also be an effective motivator for employee environmental initiatives. The environmental management literature points to this conclusion, but no previous research can be found that empirically investigates the issue. Demonstrated commitment to published environmental policy was shown by our research to have a statistically significant impact on employee eco-initiatives, and an interesting and never analyzed before indirect effect.

Eleven of the other twelve environmental sub-policies we tested in the survey did not have a statistically significant direct impact (with the exception of fossil fuel reduction policies having a significant negative impact). These eleven policies are often embedded in the written environmental policy which was found to have a statistically significant impact on eco-innovation, but when we asked about them individually they were not statistically significant. We believe that there are several possible explanations for this sub-policy result. Specific policies may have less relevance to different employees from different areas of the company. For instance, employees in the design area may be aware of the company policy on life cycle analysis, but others may not be. Or, employees may not believe that the company is committed to the policies, even if they do exist. (Note that there is a body of research on perceptions that may give some insight into why employee perceptions differ from the reality. Clearly perception differs from reality in this case as we know that the policies exist but employees perceive that the company is not committed to the policies.)

Furthermore, we believe that one of the most interesting results we found in the environmental policy area resulted from the interaction between policy and behaviors independent variables. Specifically, we found that employees who perceived that any of the policies existed were more likely to perceive support from their supervisors. Thus, the existence of environmental policies can be seen to have an indirect impact on eco-innovation as it increases the sensitivity of employees to supervisory support for environmental actions. This interaction effect which links policies to behaviors has never previously been explored in the environmental management literature. We feel this is an important contribution to the further development of theory in this area.

Third, an important conclusion from our results is that the existing environmental management theory has possibly made some incorrect assumptions as to which factors influence employee eco-innovation. Specifically, we find that environmental information dissemination does not have a statistically significant impact on employee eco-innovation, despite the consensus in the literature that it should have a positive influence. We have explored possible explanations for this lack of influence in Chapter 4. For instance, we note that supervisory behaviors that support environmental communication from employees is the factor with the greatest explanatory power on the dependent variable and therefore, we posit that how active communication from employees is received by managers may be more important to encouraging environmental innovations than passive information received from managers.

Fourth, we developed a survey questionnaire that used behaviorally-anchored rating scales (BARS) to study which organization behaviors, identified in the general organizational literature, are important for supporting employee participation in the environmental area. Our use of BARS is novel in that, to our knowledge, it has never before been used for the purpose of testing employee assessment of supervisory behaviors. Our study finds robust statistical results using this tool.

Fifth, our research shows that while supervisory support behaviors, when applied by supervisors to support the environmental activities of employees, positively influence eco-innovation, most companies are weak at encouraging these behaviors from supervisors. This finding was confirmed using a Chi-square test of differences, which showed that managers were using significantly more positive behaviors in the general management area compared to environment. And a likelihood ratio tests which confirmed that the behaviors in our BARS when applied in the environmental case are more effective at influencing employee eco-innovation than when applied to general management.

From this result we draw two conclusions. First, our results support the assumption that most of the same factors that influence other innovation also have a statistically significant impact on eco-innovation, although not all have as strong an impact. The exception of information dissemination support behaviors has been noted. Second, employees from companies in our sample, which is representative of large environmentally- proactive firms head-quartered in Europe with international business operations, did not have supervisors who used equally supportive behaviors when managing employee environmental activities as compared to other activities. From this second conclusion we note that most companies probably are not effectively encouraging supervisors to use the positive behaviors when managing environmental actions, since even those companies who have a strong commitment to do so have not yet successfully done so.

5.2   IMPLICATIONS FOR RESEARCH AND PRACTICE

This research has important implications for both researchers and practitioners. This section first describes the implications for theory, then offers practical implications for managers in corporations.

Implications: Organizational Behavior Research

The organizational behavior literature of Cummings (1965) and Lawyer (1992) indicates that control-oriented management, defined as hierarchical and bureaucratic in nature, prevents the full participation of employees in innovation. High involvement or participative management, defined as non-hierarchical, non-bureaucratic, and open/inviting of employee input into the decision-making processes, leads to empowered employees who tend to find innovative business solutions (Bandura, 1977; Conger & Kanungo, 1988; Lawyer, 1986, 1992; Kanter, 1983; Spreitzer, 1995). Our research provides an empirical investigation that studies whether the types of behaviors found in "high-involvement organizations" support employee initiatives. This literature indicates the importance of supervisory behaviors that include openness of communication, goal-setting, non-bureaucratic and non-hierarchical approaches, openness to employee participation in decision-making, feedback on performance and goals, openness to experimentation, and exposure of employees to learning opportunities. Our dissertation research tested for these types of support behaviors. And, the results of our empirical study show that these types of supportive behaviors on the part of supervisors do increase the probability of employee initiatives. We were looking at the special case of eco-initiatives, but the general conclusion that high-involvement management styles encourage innovation can also be made from these results. (Note that the types of organizations in our sample may have been more likely to have supportive behaviors than the norm, as we chose a sample of proactive companies who might tend to have more supportive management.)

Lawyer (1986, 1990, 1992) talks about the importance of incentives in encouraging employee participation. We tested whether employees were more likely to try eco-initiatives if supervisors used rewards and recognition for eco-responsible behavior, and we found there was a significant positive relationship. Thus, the conclusions of the general literature on incentives was confirmed to apply to the specific case of environmental management.

We found one exception in our work to the suppositions of the empowerment literature. Both Lawyer (1992) and Spreitzer (1995) note that information dissemination promotes employee empowerment and involvement in innovations. But, our research does not bear this point out. While the communication of a written environmental policy, which shows the company environmental strategy and vision, appears to have a clear positive impact on employee initiatives, communication of other types of environmental information does not seem to have a clear impact. As noted above, we feel this is an area that could benefit from a further exploration in order to discover which types of information sharing are important to encourage both innovation in general and environmental innovation on the part of employees. For instance, in the environmental case, is employee innovation encouraged when specific information on progress toward company targets is disseminated by managers?

Research on organizational learning and innovation (Argyris and Schon, 1978; Cavaleri and Fearon, 1996; Luthans and Kreitner, 1985; Peters, 1991; Redmond, Mumford, and Teach, 1993; Senge, 1990) indicates that supervisors have an important role in encouraging employee innovations. And, Van de Ven (1986) notes that lack of management support acts as a barrier to innovation in organizations. Innovation occurs in organizations where individuals and groups undergo a search for new ways of doing things. This search process is by its nature a learning process for employees in the organization. Employees must have an impetus to break with past routines and search for new and better ways of performing in the work place. Supervisors play an important role in demonstrating organizational commitment to the search for better processes and provide daily feedback as to the desirability and direction of the search for innovative solutions.

Therefore, the organizational behavior and innovation literature asserts that supervisors can help employees learn that a different set of responses are necessary and that the organization will support employee activities that involve searching for better business responses. Supervisors provide signals as to the organization's and their own commitment to organizational change and to this learning process by the way they behave when managing daily business activities. Our research shows that employees are very savvy about picking up the signals given by supervisory behaviors. Specifically, our research shows a systematic linkage between supervisory behaviors that support change and innovation and employees' likelihood of taking the risk to try to promote a new way of doing things. Thus, we provide empirical evidence that employees with supervisors who are supportive of learning and innovation behave more proactively and creatively than those who do not believe (based on the daily behavioral signals from their supervisor) management of the company desires that change. The implications of this general conclusion may be taken one step further, i.e. employees may be more likely to learn new behaviors that help the business (e.g. come up with innovative solutions) if their supervisors have behaviors that support learning and are capable of learning themselves. While this has not been the subject of our empirical investigation, it could be an interesting area for future research.[1] Specifically, research could be conducted in companies to find out if the learning behaviors that are described in the BARS are more likely to be demonstrated by managers whose employees try to promote innovations in general.

Our derived model of factors affecting employee willingness to innovate (Figure 1.1) notes the areas which the literature finds important in supporting employee activities. We have incorporated these factors into our conceptual model (Figure 2.1), showing through our empirical investigation that these factors also have an important impact on employee environmental initiatives. (As noted, only one factor from our model did not have a statistically significant impact. That factor was information dissemination.) Thus, we have made linkages between the organizational behavior theory and environmental management.

Finally, we do not know of any literature that has looked at the interaction between policies and supervisory support behaviors' impacts. We empirically find a conditional effect that could lead to further theoretical work in this direction.

Thus, we see that in general our work supports and complements existing theory of organizational behavior, except in the case of information dissemination. Specifically, our study offers empirical supports that behaviors which support innovative actions by employees, employee competence building, open communication and non-hierarchical structures, use of rewards and recognition, and management of goals and responsibilities are important for employee innovations. But, our research also allows the differentiation between factors, indicates that information dissemination can not be broadly assumed to have an impact on employee actions as the literature indicates.

Implications: Environmental Management Research

Our study contributes significantly to the area of research which Stern (1992) describes as human environmental attitudes and behaviors in firms. Our findings support the assumption that the behaviors of supervisors do indeed impact the environmental performance of a firm in so far as they influence employee actions. Our research also supports the work of Shrivastava and Hart (1994) which asserts that the greening of corporations requires a new way of managing people. In agreement with Post and Altman (1992) who assert that organizational learning can help firms manage activities that affect the natural environment, our conceptual model uses organizational learning as a theoretical basis. And, our empirical work, which showed a positive correlation between supervisors' use of supportive behaviors (as are found in learning organizations) and employee environmental innovation, supports the literature that assumes organizational learning will help companies become more sustainable.

Management commitment, as indicated by Davis (1991), Friedman (1992), Hunt and Auster (1990), Rappaport and Dillon (1991) and Zeffane, Polonsky, and Metley (1995), influences organizational and supervisory support for environmental protection activities in a firm. Top management communication of the environmental policy is one way firms can successfully persuade employees of organizational commitment. While our empirical work does not try to ascertain who in the firm communicates the environmental policy, we do have strong evidence that communication of such a policy positively influences employee participation in promoting eco-initiatives. Thus, our investigation confirms the importance given in the literature to top management commitment to and communication of policy.

As Hostager, Neil, Decker, and Lorentz (1998) note, the intrinsic motivation of employees who care about environmental protection can affect their willingness to eco-innovate. This intrinsic motivation can explain the likelihood that some respondents in the survey, who tried environmental initiatives, did so despite a lack of perceived support on the part of their supervisors and/or organization. Yet, organizational support still had a statistically significant influence on employee eco-innovations. (Note that our survey did not allow us to determine which respondents were intrinsically motivated to protect the environment as it was an issue that was outside the purview of our study. Future research into the area of personal environmental commitment to environmental protection would be of interest to explore the tradeoffs between organizational support and personal values.)

It is possible to draw a further conclusion with regards to the similarities between support for innovation in general and eco-innovation. Van de Ven (1986) notes that the absence of managerial attention is a barrier to innovation in organizations. We found that in the absence of managerial attention to environmental management, employees were less likely to eco-innovate. This is an expected result, but none the less worth mentioning. Our results lead us to conclude that motivating environmental innovations from employees may be even more difficult than motivating innovations in general as environmental protection is perceived by many employees and managers as outside the raison d'être of the firm.

Implications for Practice

Here we discuss possible implications from our research for businesses that want to harness employee ideas in order to increase the sustainability of their activities.

Since we find that supervisory behaviors that support environmental actions have an important influence on employees eco-initiatives, organizations that want to increase employee eco-innovation may do well to encourage managerial support of such initiatives using the positive behaviors listed in Table 3.1. Changing behaviors of management personnel to those that are more supportive of employee innovations is indicated. For example, encouraging managers at all levels of the company to practice openness to experimentation and the search for environmental solutions, to support environmental competence building of employees including site visits and assignment rotations that encourage learning, to have open styles of communication that emphasize employee participation in decision-making and non-hierarchical processes, to use formal rewards and informal recognition to reinforce individual and organizational environmental goals, and to manage environmental objectives by sharing responsibility with each employee for environmental performance and solution finding. Change in supervisory behaviors will need organizational support of not only a clear commitment to environmental policy, but time and resource allocation as well. One possible approach would be to provide supervisors with management training and competence building in this area. Another approach would be use of a mentoring program where supportive managers act as role models for those who have been shown to be using less supportive behaviors. Our results show which support behaviors should be encouraged. Our results provide a ranked list showing environmental communication and competence building as the most efficient at increasing eco-innovation, and then rewards and recognition of environmental activities, management of environmental goals and responsibilities and support for environmental innovation.

Note also that while some supervisors may already be practicing supportive behaviors, such as those listed above, our findings show that they may only be doing so in the general management case. Our empirical results show a stronger impact when supervisors use these behaviors to manage environmental activities. Thus, the organization may need to support a shift toward integrating environmental management considerations into general business decision-making, encouraging the use of these behaviors in the process, if they want to increase employee eco-innovation. And, companies who want eco-innovation need to focus concerted effort on removing these barriers of managerial neglect if they want to encourage eco-innovation.

Businesses may also want to note that often employees are missing the capacity to participate fully in the search for environmentally-friendly business solutions. While the results of the survey indicated that support for environmental competence building had a significant positive impact on the probability of employee initiatives (the most supportive behavior gave a probability of 64 percent that an employee would have tried to promote an eco-initiative versus 30 percent for an employee whose supervisor gave the least supportive behavior), companies appear to be weak in providing environmental education. We conclude that companies may need to emphasize environmental skills and knowledge development in order to benefit from employee eco-innovations and provide here the tool for a differentiated and organized administration of this emphasis.

Communication of commitment to environmental policy was shown to have a strong impact on employee eco-innovation. Managers can learn from this result that focusing time and energy on communicating the corporate vision (policy) of sustainable development can empower employees to search for environmental solutions. Thus, if a company wants to move toward sustainability, it should consider focusing energy on the development of an environmental vision and the communication of this policy as part of the core business activities of all employees.

Note that all of the companies in the survey had most of the environmental policies for which we tested, but that employee perceptions of company commitment to these policies was often weak. Companies who have the listed environmental policies, may not in reality be implementing these policies, since employees perceive a lack of corporate commitment. Bridging the gap between the stated intention of protecting the natural environment and conducting business in an environmentally sustainable manner and actually doing so will, in our view, require greater efforts.

Specifically, line management can increase employee participation by adopting behaviors that encourage environmental actions, thus changing the perceptions of these employees and increasing the sustainability of the firm.

5.3   FUTURE RESEARCH

One weakness of our research is its reliance on self-reported eco-initiatives. While the definition of our dependent variable allowed us to test organizational policy and supervisory behavior independent variables, and was sufficiently robust for our work, narrowing the definition of an initiative to include only eco-innovations that improve the sustainability of an organization, could be an interesting refinement. There are obvious difficulties in communicating such a "sustainable eco-innovation" dependent variable. As such, independently verifying (possibly using a separate survey tool or interviews) what these initiatives were in practice, might strengthen the results of future studies.

We selected a sample of companies with proactive environmental policies and expected that they would have supervisors who provided above average support for employee environmental innovations. A future survey in companies without a bias toward a support for eco-innovations could reveal some interesting data for comparison.

Also enlarging the current sample by broadening it to include employees from more companies operating in other countries would allow us to further test our results and may help to increase confidence in the ranked list of significant behaviors. This additional research would allow us to test whether the current ranking of significant results is robust.

With regards to our specific findings, we note that research into the effect of information dissemination on eco-innovations would be of interest. Supervisors may not be relating environmental information, such as progress toward company goals or environmental activities occurring elsewhere in the company, in a way that employees can relate this information to their personal actions. For instance, information on company targets may not seem relevant in its aggregate form to personal employee targets. This area would warrant further research.

In the area of environmental policies there are two possible open questions deserving exploration. First, since eleven of the thirteen policies were not found to have a statistically significant impact on employee eco-innovation it would be interesting to study the question of why. For instance, are these policies less relevant to employees' daily work? Are they less well communicated by the companies than other policies? Are companies less committed to these specific policies? Or are there other alternative explanations for this result.

Secondly, future research to better understand the interaction between policies and supervisory behaviors could lead to greater clarity as to how these affect eco-innovation, and probably company performance in general. While our research shows a definite interaction between these variables, this interaction is complex in nature. Exploring further the nature of the relationship could have some interesting implications for environmental management theory and beyond.

Finally, we note that employees and managers have pre-existing values and capabilities that can affect eco-innovation. It would be interesting to see the results of a study which controls for these personal values and capacity factors.

***

In conclusion, we believe our survey design successfully tested our hypotheses. And, we feel that the survey questionnaire was an effective tool for gathering data about employee perceptions of organizational and supervisory support for employee eco-initiatives. Specifically, the use of BARS instead of Likert scale opinion survey methods provided a useful set of data from which to draw conclusions.

Our empirical findings demonstrate the contribution we have made to measuring the differentiated impact of factors affecting employee eco-initiatives. Since there has been an absence of empirical work in this area, we hope that this contribution will lead to further research initiatives to address some of the unanswered questions that have come from this investigation. If nothing else, we hope that our research incites companies to take a more proactive approach to developing environmental policies and encouragement for line management and employees to participate fully in sustainable business practices, supporting the search of innovative solutions to ecological problems.

References

Abbey, A. & Dickson, J. W. 1983. R & D climate and innovation in semiconductors. *Academy of Management Journal*, 26: 362-368.

Amabile, T. M. 1983. The social psychology of creativity: A componential conceptualization. *Journal of Personality and Social Psychology*, 45: 357-376.

Amabile, T. M. 1988. A model of creativity and innovation in organizations. In B. M. Staw & L. L. Cummings (Eds.). *Research in Organizational Behavior*, 10: 123-167. Greenwich, CT: JAI Press.

Amabile, T. M. 1993. Motivational synergy: Toward new conceptualizations of intrinsic and extrinsic motivation in the workplace. *Human Resource Management Review*, 3: 185-201.

Amabile, T. M., Conti, R., Coon, H., Lazenby, J. & Herron, M. 1996. Assessing the work environment for creativity. *Academy of Management Journal*, 39(5): 1154-1184.

Andrews, F. M. 1975. Social and psychological factors that influence the creative process. In I. A. Taylor & J. W. Getzels (Eds.). *Perspectives in creativity*, 117-145. Chicago: Aldine.

Argyris, C. & Schon, D. 1978. *Organizational Learning: A Theory of Action Perspective*. Reading, MA: Addison-Wesley.

Arthur D. Little 1988. Environmental, health and safety policies: current practices and future trends. *Arthur D. Little, Inc. Center for Environmental Assurance*, Cambridge, MA.

Arthur D. Little 1989. State of the art environmental, health and safety management programs: how do you compare? *Arthur D. Little, Inc. Center for Environmental Assurance*, Cambridge, MA.

Ayres, R. 1995. Achieving eco-efficiency in business. *Report of World Business Council for Sustainable Development Second Antwerp Eco-Efficiency Workshop*, March.

Bandura, A. 1977. Self-efficacy: Toward a unifying theory of behavioral change. *Psychological Review*, 84: 191-215.

Bandura, A. 1986. *Social Foundations Of Thought And Action: A Social Cognitive Theory*. Englewood Cliffs, NJ: Prentice-Hall.

Barrett, S. & Murphy, D. 1996. Managing corporate environmental policy: A process of complex change. In Wehrmeyer, W. (Ed.). *Greening People*, 75-98. Sheffield, England: Greenleaf Publishing.

Bass, B. M. 1981. *Bass and Stodghill's Handbook of Leadership*. NY: Free Press.

BCSD 1994. *Internalizing environmental costs to promote eco-efficiency*. Tomorrow Publishing, Stockholm, Sweden.

Beard, C. & Hartmann, R. 1997. Naturally enterprising - eco-design, creative thinking and the greening of business products. *European Business Review*, 97 (5): 237-243.

Becker, T. E., Billings, R. S., Eveleth, D. M. & Gilbert, N. L. 1996. Foci and bases of employee commitment: implications for job performance. *Academy of Management Journal*, 39(2): 464-482.

Belz, F. et al. 1997. From the eco-niche to the ecological mass-market in the food industry. Discussion paper No. 40 of the *Institute for Economy and the Environment at the University of St. Gallen*, St. Gallen 1997.

Belz, F. & Strannegard, L. 1997. *International business environmental barometer*. Oslo: Cappelen Akademisk Forlag.

Bennis, W. & Nanus, B. 1985. *Leaders*. NY: Harper and Row.

Bird, B. J. 1989. *Entrepreneurial Behavior*. Glenview, IL: Scott, Foresman.

Bluhdorn, I., Krause, F. & Scharf, T. (Eds.). 1995. *The green agenda: environmental politics and policies in Germany*. Keele, Staffordshire, UK: Keele University Press.

Bowen, D. E. & Lawler, E. E. 1992. The empowerment of service workers: What, why, how and when. *Sloan Management Review*, Spring: 31-39.

Brazeal, D. V. 1993. Organizing for internally developed corporate ventures. *Journal of Business Venturing*, 8(1): 75-90.

Brazeal, D. V. 1996. Managing an entrepreneurial organizational environment: a discriminant analysis of organizational and individual differences between autonomous unit managers and department managers. *Journal of Business Research*, 35(1): 55-67.

Bringer, R. P. & Benforado, D. M. 1993. Industry response to waste challenge. *Forum for Applied Research and Public Policy* (Spring) 8(1): 60-70.

Brophy, M. 1996. The essential characteristics of an environmental policy, and environmental guidelines and charters. Both in Welford, R. (Ed.). *Corporate Environmental Management: Systems and Strategies*, 92-103, 104-118. London: Earthscan.

Buchanan, D. A. & McCalman, J. 1987. Micro 11 survey, 1987: Digital Equipment Scotland. *Centre for technical and organizational change*, company confidential report, Glasgow, June.

Buchanan, D. A. & McCalman, J. 1989. *High performance work design: the digital experience*. London: Routledge.

Burgelman, R. A. 1983. A process model of internal corporate venturing in the diversified major firm. *Administrative Science Quarterly*, 28: 223-244.

Campbell, T. & Cairns, H. 1994. Developing and measuring the learning organization: From buzz words to behaviors. *Industrial and Commercial Training*, 26 (7): 10-15.

Carpenter, S. R. 1993. When are technologies sustainable? In L. Hickman & C. Porter (Eds.). *Technology and Ecology*: 202-214. Carbondale, IL.: Society for Philosophy and Technology.

Carson, P. & Moulden, J. 1991. Green is gold. *Small Business Reports*, 16 (12): 68-71.

Carson, R. 1962. *Silent Spring*. Cambridge, MA: The Riverside Press.

Cavaleri, S. & Fearon, D. 1996. *Managing in organizations that learn*. Oxford: Blackwell Business.

CERES 1995. Guide to the CERES Principles. *Coalition for environmentally responsible economies.* Boston.

Colburn, T., Dumanoski, D. & Myers, J. P. 1996. *Our Stolen Future.* New York: Penguin - Dutton.

Colby, J. 1991. Environmental management in development: the evolution of paradigms. *Ecological Economics,* 3: 291-313.

Conger, J. A. & Kanungo, R. N. 1988. The empowerment process: Integrating theory and practice. *Academy of Management Review,* 13(3): 471-482.

Cooper, D. R. & Emory, W. C. 1991. *Business Research Methods.* Chicago: Irwin.

Cramer, J. 1996a. Experiences with implementing integrated chain management in Dutch industry. *Business Strategy and the Environment,* 5: 38-47.

Cramer, J. 1996b. Instruments and strategies to improve the eco-efficiency of products. *Environmental Quality Management,* 6 (2): 57-65.

Cramer, J. S. 1991. *The logit model.* NY: Chapman and Hall.

Cummings, L. L. 1965. Organizational climates for creativity. *Journal of the Academy of Management,* 3: 220-227.

Datta, K. 1995. Measuring environmental performance. *Environmental Protection,* August: 39.

Davis, D. 1986. Technological innovation and organizational change. In D. Davis (Ed.). *Managing Technological Innovation,* 1-16. San Francisco, CA: Jossey-Bass.

Davis, J. 1991. *Greening business: Managing for sustainable development.* Oxford: Basil Blackwell.

Delbecq, A. L. & Mills, P. K. 1985. Managerial practices that enhance innovation. *Organizational Dynamics,* 14(1): 24-34.

Doherty, B. & Rawcliffe, P. 1995. British exceptionalism? Comparing the environmental movement in Britain and Germany. In Bluhdorn, I., Krause, F., & Scharf, T. *The Green Agenda: Environmental Politics And Policy In Germany,* 235-250. Keele, Staffordshire, UK: Keele University Press.

Dow Annual Report to Shareholders 1972.

Ehrenfeld, D. 1978. *The arrogance of humanism.* New York: Oxford University Press.

Ellison, R. L., James, L. R., McDonald, B. W., Fox, D. G. & Taylor, C. W. 1968. Goals: An approach to motivation and achievement. *Journal of Personality and Social Psychology,* 54: 5-12.

Epstein, M. 1996. *Measuring corporate environmental performance: Best practices for costing and managing an effective environmental strategy.* Chicago, IL: Irwin.

Ettlie, J. E. 1983. Organizational policy and innovation among suppliers to the food processing sector. *Academy of Management Journal,* 26: 27-44.

Farr, J. L., Enscore, E. E., Dubin, S. S., Cleveland, J. N. & Kozlowski, S. W. J. 1980. Behaviorally Anchored Scales - A Method for Identifying Continuing Education Needs of Engineers. *The Pennsylvania State University,* Final Report.

Fischer, K. & Schot, J. 1993. *Environmental Strategies for Industry: International Perspectives on Research Needs and Policy Implications.* Washington, D. C.: Island Press.

Flaherty, M. & Rappaport, A. 1991. *Multinational Corporations and the Environment: A Survey of Global Practices.* Medford, MA.: Center for Environmental Management, Tufts University.

Frankel, C. 1995. The visions gap. *Tomorrow* (July-September) 5(3): 72-74.

Friedman, F. B. 1992. *Practical guide to environmental management.* Washington, D.C.: The Environmental Law Institute.

Fry, A. 1987. The post-it note: an intrapreneurial success. *SAM Advanced Management Journal,* 52(3): 4-9.

Fussler, C. 1996. *Driving eco-innovation: A breakthrough discipline for innovation and sustainability.* London: Pitman Publishing.

Garvin, D. A. 1993. Building a Learning Organization. *Harvard Business Review,* July-August: 78-92.

Gladwin, T. 1977. Environmental policy trends facing multinationals. *California Management Review,* 20 (2): 81-93.

Gladwin, T. 1992. *Building the Sustainable Corporation: Creating Environmental Sustainability and Competitive Advantage.* Washington, D.C.: National Wildlife Federation.

Gladwin, T. & Welles, J. G. 1976. Multinational corporations and environmental protection: patterns of organizational adaptation. *International Studies of Management and Organization,* 9: 160-184.

Gladwin, T. N., Kennelly, J. J. & Krause, T. S. 1995. Shifting paradigms for sustainable development: implications for management theory and research. *Academy of Management Review,* 20 (4): 874-907.

Goitein, B. 1989. Organizational decision-making and energy conservation in investments. *Evaluation Program Planning,* 12:143-51.

Gray, R., Bebbington, J. & Walters, D. 1993. *Accounting for the environment: The greening of accountancy part II,* London: Paul Chapman Publishing.

Green, K., Morton, B. & New, S. 1995. Environmental impact of purchasing in organizations. *Paper presented to Greening of Industry Conference,* Toronto.

Greene, W. H. 1993. *Econometric analysis.* Second Edition, Englewood Cliffs, NJ: Prentice-Hall.

Grossman, S. R. 1982. Training creativity and problem-solving. *Training and Development Journal,* 36: 62-68.

Group of Lisbon, 1995. *Limits to Competition,* Cambridge, MA: MIT.

Hage, J. & Dewar, R. 1973. Elite values versus organizational structure in predicting innovation. *Administrative Science,* 18: 279-290.

Hampton, D., Summer, C. & Webber, R. 1987. *Organizational behavior and the practice of management.* Glenview, IL: Scott, Foresman.

Harrison, R. 1983. Strategies for a new age. *Human Resource Management,* 22: 209-235.

Hart, S. 1997. Beyond greening: strategies for a sustainable world. *Harvard Business Review,* January-February: 66-77.

Hawken, P. 1993. *The ecology of commerce*. New York: Harper Collins.

Heaton, G., Repetto, R. & Sobin, R. 1991. *Transforming technology: An agenda for environmentally sustainable growth in the 21st century*. Washington, D.C.: World Resources Institute.

Herzberg, F. 1982. *The managerial choice: To be efficient and to be human*. Rev. ed. Salt Lake City: Olympus.

Hoffman, A. 1997. *From heresy to dogma: An institutional history of corporate environmentalism*. San Francisco: The New Lexington Press.

Hostager, T. J., Neil, T. C., Decker, R. L. & Lorentz, R. D. 1998. Seeing environmental opportunities: effects of intrapreneurial ability, efficacy, motivation, and desirability. *Journal of Organizational Change Management*, 11(1): 11-25.

House, R. J. & Mitchell, T. R. 1968. Path goal theory of leadership. *Journal of Contemporary Business*, 3: 81-97.

Huczynski, A. & Buchanan, D. 1991. *Organizational behavior: An introductory text*. Englewood Cliffs, NJ: Prentice-Hall.

The Human Environment 1972. Proceedings report from the United Nations Conference on the Human Environment. Stockholm.

Hunt, C. B. & Auster, E. R. 1990. Proactive environmental management: avoiding the toxic trap. *Sloan Management Review*, Winter: 7-18.

Hutchinson, A. & Hutchinson, F. 1995. Sustainable regeneration of the UK's small and medium-sized enterprise sector: some implications of SME response to BS7750. *Greener Management International*, 9: 73-84.

Hutchinson, C. 1996. Corporate strategy and the environment. In Welford, R. & Starkey, R. (Eds.). *Business and the Environment, 85-104*. London: Earthscan.

Isaksen, S. G. 1983. Toward a model for the facilitation of creative problem solving. *Journal of Creative Behavior*, 17: 18-31.

Jaques, E. 1977. *A general theory of bureaucracy*. London: Heinemann.

James, L. R. & James, L. A. 1989. Integrating work environment perceptions: Explorations into the measurement of meaning. *Journal of Applied Psychology*, 74: 739-751.

Kanter, R. M. 1979. Power failure in management circuits. *Harvard Business Review*, 57 (4): 65-75.

Kanter, R. M. 1983. *The change masters*. New York: Simon & Schuster.

Kanter, R. M. 1989a. The new managerial work. *Harvard Business Review*, 89(6): 85-92.

Kanter, R. M. 1989b. *When giants learn to dance*. New York: Simon & Schuster.

Kelley, G. 1976. Seducing the elites: the politics of decision making and innovation in organizational networks. *Academy of Management Review*, 1(3): 66-74.

Kemp, R. & Soete, L. 1992. The greening of technological progress: an evolutionary perspective. MERIT Research Memorandum 91-011. *Futures*, 24 (5): 437-57.

Keogh, P. D. & Polonsky, M. J. 1998. Environmental commitment: a basis for environmental entrepreneurship? *Journal of Organizational Change Management*, 11(1): 38-49.

Kiernan, M. J. & Levinson, J. 1997. Environment drives financial performance: The jury is in. *Environmental Quality Management*, Winter: 1-7.

Kimberley, J. R. & Evanisko, M. J. 1981. Organizational innovation: The influence of individual, organizational, and contextual factors on hospital adoption of technological and administrative innovations. *Academy of Management Journal*, 24: 689-713.

Kinlaw, D. C. 1993. *Competitive and green: Sustainable performance in the environmental age*. Amsterdam: Pfieffer.

Kuratko, D. F., Montagno, R. V. & Hornsby, J. S. 1990. Developing an intrapreneurial assessment instrument for an effective corporate entrepreneurial environment. *Strategic Management Journal*, 11: 49-58.

Landy, F. & Farr, J. 1983. *The measurement of work performance: Method, theory & applications*. NY: Academic Press.

Landy, M., Roberts, M. & Thomas, S. 1990. *The environmental protection agency: Asking the wrong questions*. New York: Oxford University Press.

Langley, P. & Jones, R. 1988. A computational model of scientific insight. In R. J. Sternberg (Ed.). *The nature of creativity*: 177-201. NY: Cambridge University Press.

Lawler, E. E. 1973. *Motivation in work organizations*. Pacific Grove, CA: Brooks and Cole.

Lawler, E. E. 1986. *High-involvement management*. San Francisco: Jossey-Bass.

Lawler, E. E. 1990. *Strategic pay: Aligning organizational strategies and pay systems*. San Francisco: Jossey-Bass.

Lawler, E. E. 1992. *The ultimate advantage: Creating the high involvement organization*. San Francisco, Jossey-Bass.

Lefebvre, L., Lefebvre, E. & Roy, M-J. 1995. Integrating environmental issues into corporate strategy: a catalyst for radical organizational innovation. *Creativity and Innovation Management*, 4(4): 209-222.

Levitt, B. & March, J. G. 1988. Organizational learning. *Annual Review of Sociology*, 14: 319-40.

Likert, R. 1961. *New patterns of management*, NY: McGraw-Hill.

Lober, D., Bynum, D., Campbell, E. & Jacques, M. 1997. The 100 plus corporate environmental report study: a survey of an evolving environmental management tool. *Business Strategy and the Environment*, 6: 57-73.

Luthans, F. & Kreitner, R. 1985. *Organizational behavior modification and beyond*. Glenview, IL: Scott, Foresman.

Magretta, J. 1997. Growth through global sustainability: an interview with Monsanto's CEO, Robert B. Shapiro. *Harvard Business Review*, January-February: 78-90.

Makower, J. 1993. *The E factor: The bottom-line approach to environmentally responsible business*. NY: Random House Tilden Press.

March, J. G. & Simon, H. A. 1958. *Organizations*. NY: John Wiley & Sons.

Maxwell, J., Rothenberg, S., Briscoe, F. & Marcus, A. 1997. Green schemes: corporate environmental strategies and their implementation. *California Management Review*, 39(3): 118-134.

McClelland, D. C. 1975. *Power: The inner experience.* NY: Irvington Press.

McLaughlin, A. 1993. *Regarding nature.* Albany: State University of New York Press.

Meadows, D. H., Meadows, D. L. & Randers, J. 1992. *Beyond the limits.* Post Mills, VT: Chelsea Green.

Meadows, D. H., Meadows, D. L., Randers, J. & Behrens III, W. W., 1972. *The limits to growth: A report of the Club of Rome.* NY: Universe Books.

Meffert, H., Brenkenstein, M. & Schubert, F. 1987. Umweltschutz und unternehmensverhalten. *Harvard Manager* (2): 32-39.

Merton, R. K. 1936. The unanticipated consequences of purposive social action. *American Sociological Review*, 1: 894-904.

Merton, R. K. 1940. Bureaucratic structure and personality. *Social Forces*, 18: 560-568.

Merton, R. K. 1945. Role of intellectual in public bureaucracy. *Social Forces*, 23: 405-415.

Merton, R. K. 1947. The machine, the worker and the engineer. *Science*, 105: 79-84.

Merton, R. K. 1957. *Social theory and social structure.* Glencoe, IL: Free Press.

Milbrath, L. W. 1989. *Envisioning A Sustainable Society.* Albany: State University of New York Press.

Milliman, J. & Clair, J. 1995. Environmental HRM best practices in the USA: a review of the literature. *Greener Management International*, 10: 34-48.

Mirvis, P. 1996. Environmentalism in progressive businesses. *Journal of Organizational Change Management*, 7(4): 82-100.

Mitchell, T. R. 1978. *People in organizations: Understanding their behavior.* NY: McGraw-Hill.

Mumford, M. D. & Gustafson, S. B. 1988. Creativity syndrome: Integration, application, and innovation. *Psychological Bulletin*, 103: 27-43.

Netherwood, A. 1996. Environmental management systems. In Welford (Ed.). *Corporate environmental management: systems and strategies, 35-58.* London: Earthscan.

Nijkamp, P. & Soeteman, F. 1988. Ecologically sustainable economic development: key issues for strategic environmental management. *International Journal of Socioeconomics*, 15(3-4): 88-102.

Park, J. 1997. Business and environment: a reluctant partnership. *Ecodecision*, Spring: 30-32.

Pasmore, W. A. 1994. *Creating strategic change: Designing the flexible, high-performing organization.* NY: John Wiley.

Pearn, M., Roderick, C. & Mulrooney, C. 1995. *Learning organizations in practice.* NY: McGraw-Hill.

Pelz, D. C. 1956. Some social factors related to performance in research organizations. *Administrative Science Quarterly*, 1: 310-325.

Peters, T. 1991. Get innovative or get dead (part 1) and The open corporation (part 2), *California Management Review*, Fall (1990): 9-26 and Winter (1991): 9-23.

Peters, T. & Waterman, R. 1982. *In search of excellence.* NY: Harper and Row.

Petulla, J. 1987. Environmental management in industry. *Journal of Professional Issues in Engineering*, 113(2): 167-83.

Pezzey, J. 1989. *Economics analysis of sustainable growth and sustainable development.* Working paper, Environment Department, World Bank, Washington, D. C.

Pinchot, G. 1985. *Intrapreneuring.* NY: Harper & Row.

Pinchot, G. & Pinchot, E. 1994. *The intelligent organization: Engaging the talent and initiative of everyone in the workplace.* San Francisco: Berrett-Koehler.

Polonsky, M., Zeffane, R. & Medley, P. 1992. Corporate environmental commitment in Australia: a sectorial approach. *Business Strategy and the Environment*, 1(2): 25-40.

Porter, M. 1980. *Competitive strategy.* NY: The Free Press.

Post, J. & Altman, B. 1992. Models of corporate greening: how corporate social policy and organizational learning inform leading-edge environmental management. In Post, J. (Ed.). *Research in Corporate Social Performance and Policy*, 13: 3-29. Greenwich, CT: JAI Press.

Post, J. & Altman, B. 1994. Managing the environmental change process: barriers and opportunities. *Journal of Organizational Change Management*, 7(4): 64-81.

Post, J. E. 1991. Managing as if the earth mattered. *Business Horizons*, 34(4): 32-38.

Quinn, R. E. & Spreitzer, G. M. 1997. The road to empowerment: Seven questions every leader should consider. *Organizational Dynamics*, Autumn: 37-49.

Raffee, H. 1979. *Marketing und umwelt*, Stuttgart.

Ramus, C. A., Steger, U. & Winter, M. 1996. Environmental protection can give a competitive edge. *Perspectives for Managers*, Lausanne, Switzerland: IMD.

Randolph, W. A. 1995. Navigating the Journey to Empowerment. *Organizational Dynamics*, Spring: 19-32.

Rands, G. P. 1990. Environmental attitudes, behaviors, and decision making: implications for management education and development. In Hoffman, W. M., Frederick, R. & Petry, E. S. (Eds.). *The Corporation, Ethics, and the Environment*, 269-286. Westport, CT: Quorum Books.

Rappaport, A. & Dillon, P. 1991. Private-sector environmental decision making. In R. A. Chechile & S. Carlisle (Eds.). *Environmental decision making: A multi-disciplinary perspective*, 238-268. NY: Van Nostrand Reinhold.

Rappaport, A., Taylor, G. J., Flaherty, M. & Pomeroy, G. 1991. *Global corporate environmental, health, and safety programs: Management principles and practices.* Medford, MA.: Center for Environmental Management, Tufts University.

Rasanen, K., Merilainen, S. & Lovio, R. 1995. Pioneering descriptions of corporate greening: notes and doubts on the emerging discussion. *Business Strategy and the Environment*, 3(4): 9-16.

Redmond, M. R., Mumford, M. D. & Teach, R. 1993. Putting creativity to work: effects of leader behavior on subordinate creativity. *Organizational Behavior And Human Decision Processes*, 55: 120-151.

Robert, K. H. 1989. *The "Natural Step" to a sustainable future*, Stockholm: The Natural Step.

Roddick, A. 1995. Corporate responsibility: the Body Shop choice. Focus on management: creating the future. *EFMD Forum*: February: 28-33.

Rodgers, C. 1979. *Freedom to learn*. Columbus, OH: Charles E. Merrill.

Roome, N. 1992. Developing environmental management strategies. *Business Strategy and the Environment*, 1(1): 11-24.

Schaltegger, S. 1997. Economics of life cycle assessment: inefficiency of the present approach. *Business Strategy and the Environment*, 6: 1-8.

Schaltegger, S. & Sturm, A. 1990. Ecological rationality. *Die Unternehmung*, 4: 273-290.

Schaltegger, S. & Sturm, A. 1991. Okologieorientiertes management. In Frey, R. L., Staehelin-Witt, E. & Blochinger, H. (Eds.). Mit okonomie zur okologie: Analyse und losungen des umweltproblems aus okonomischer sicht, 269-300. Basel.

Schmidheiny, S. 1992. *Changing course: A global business perspective on development and the environment*. Cambridge, MA: MIT Press.

Schot, J. 1992. Credibility and markets as greening forces in the chemical industry. *Business Strategy and the Environment*, 1(1): 35-44.

Schot, J., De Laat, B., Van der Meijden, R. & Bosma, H. 1991. *Caring for the environment. Environmental management in the chemical industry*. The Hague: SDU.

Schuler, R. S. 1986. Fostering and facilitating entrepreneurship in organizations: implications for organizational structure and human resource management practices. *Human Resource Management*, 25(4): 607-29.

Schumacher, E. F. 1973. *Small is Beautiful*. San Bernardino, CA: Borgo Press.

Schwab, D. P. & Heneman, H. G. 1975. Behaviorally anchored rating scales: a review of the literature. *Personnel Psychology*, 28: 549-562.

Senge, P. 1990. *The fifth discipline: The art and practice of the learning organization*. NY: Doubleday.

SETAC News 1993. *Life-Cycle Assessment*. (November) 3(6).

Sharfman, M., Ellington, R. & Meo, M. 1997. The next step in becoming "green": Life-cycle oriented environmental management. *Business Horizons*, May-June: 13-22.

Shrivastava, P. 1995a. Ecocentric management for a risk society. *Academy of Management Review*, 20 (1): 118-137.

Shrivastava, P. 1995b. Environmental technologies and competitive advantage. *Strategic Management Journal*, 16: 183-200.

Shrivastava, P. 1996. *Greening Business: Profiting the Corporation and the Environment.* Cincinnati, OH: Thomson Executive Press.

Shrivastava, P. & Hart, S. 1994. Greening organizations - 2000. *International Journal of Public Administration,* 17 (3&4): 607-635.

Smart, B. 1992. *Beyond compliance: A new industry view of the environment.* Washington, D. C.: World Resources Institute.

Smith, D. K. 1996. *Taking charge of change.* Reading, MA: Addison-Wesley.

Smith, P. C. & Kendall, L. M. 1963. Retranslation of expectations: An approach to the construction to unambiguous anchors for rating scales. *Journal of Applied Psychology,* 47: 149-155.

Spreitzer, G. M. 1995. Psychological empowerment in the workplace: Dimensions, measurement, and validation. *Academy of Management Journal,* 38 (5): 1442-1465.

Starik, M. & Rands, G. 1995. Weaving an integrated web: multilevel and multisystem perspectives of ecologically sustainable organizations. *Academy of Management Review,* 20(4): 908-935.

Starkey, R. 1996. The standardization of environmental management systems. In Welford, R. (Ed.). *Corporate environmental management: systems and strategies,* 59-91. London: Earthscan.

Stead, W. E. & Stead, J. G. 1992. *Management for a small planet: Strategic decision making and the environment.* Newbury Park, CA.: Sage.

Steger, U. 1993. The greening of the board room: How German companies are dealing with environmental issues. In Fischer, K. & Schot, J. (Eds.). *Environmental strategies for industry,* 147-166. Washington, D.C.: Island Press.

Steger, U. 1998a. *The strategic dimensions of environmental management.* London: Macmillan Press.

Steger, U. 1998b. Empirically-based evaluation and policy recommendation for the revision of EMAS. *Executive summary final report: Research Group for the Evaluation of Environmental Management Systems.* Oestrich-Winkel, Germany: Institut fur Okologie und Unternehmensfuhrung.

Stern, P. 1992. Psychological dimensions of global environmental change. *Annual Review of Psychology,* 43: 269-302.

Stern, P. C. & Aronson, E. (Eds.). 1984. *Energy use: The human dimension.* NY: W. H. Freeman and Company.

Storen, S. 1997. Sustainable product design - is there more to it than science, systems and computers? *Creativity and Innovation Management,* 6(1): 3-9.

Strebel, P. 1996. Why do employees resist change? *Harvard Business Review,* May-June: 86-94.

Strebel, P. 1998. *New compacts between leaders of change and individuals.* London: Pitman.

Sunderland, T. 1996. Environmental management standards and certification: do they add value? *Greener Management International,* 14: 28-36.

Sustainability and United Nations Environment Program 1997. Engaging stakeholders. *The Benchmark Survey: The Second International Progress*

Report on Company Environmental Reporting, United Nations Environment Program.

Sustainability Review, 1997. Electrolux and Monsanto, July. London: Sustainability.

Taylor, C. W. 1963. Variables related to creativity and productivity among men in two research laboratories. In Taylor, C. W. & Baron, R. (Eds.). *Scientific creativity: Its recognition and development*, pp. 255-271. NY: Wiley.

Taylor, F. W. 1911. *The principles of scientific management*. NY: Harper & Brothers.

Taylor, F. W. 1947. *Scientific management*. New York.

Thielemann, U. 1990. Die unternehmung als okologischer akteur? Ansatzpunkte ganzheitlicher unternehmensethischer reflexion - zur aktualitat der theorie der unternehmung Erich Gutenbergs. In Freimann, J. (Ed.). *Okologische herausforderungen der betriebswirtschaftslehre*, 43-72. Wiesbaden.

Thomas, K. W. & Velthouse, B. A. 1990. Cognitive elements of empowerment: An "interpretive" model of intrinsic task motivation. *Academy of Management Review*, 15(4): 666-681.

Torrance, E. P. 1965. *Rewarding creative behavior*. Englewood Cliffs, NJ: Prentice-Hall.

Tushman, M. & Anderson, P. 1986. Technological discontinuities and organizational environments. *Administrative Science Quarterly*, 31, 439-465.

Tushman, M. & Moore, W. (Eds.). 1988. *Readings in the management of innovation*, 2nd ed., NY: Harper Business.

Ulhoi, J. P., Madsen, H. & Rikhardsson, 1994. *Training In Environmental Management - Industry And Sustainability: Corporate Environmental And Resource Management And Educational Requirements*. Executive Summary. Aarhus, Denmark: Institut for Informationsbehandling.

Van de Ven, A. 1986. Central problems in the management of innovation. *Management Science*, 32: 590-609.

Van de Ven, A., Angle, H. & Poole, M. (Eds.). 1989. *Research on the Management of Innovation: The Minnesota Studies*. NY: Harper & Row.

Vroom, V. H. 1964. *Work and motivation*. NY: John Wiley & Sons.

Wagner, H. 1991. The open corporation. *California Management Review*, Summer: 46-60.

Walton, R. E. 1985. From control to commitment in the workplace. *Harvard Business Review*, 63(2): 77-84.

Wehrmeyer, W. 1995. Environmental management styles, corporate culture and change. *Greener Management International*, 12: 81-94.

Wehrmeyer, W. 1996. *Greening people*. Sheffield, England: Greenleaf Publishing.

Wehrmeyer, W. & Parker, K. 1995. Identification, analysis, and relevance of environmental corporate cultures. *Business Strategy and the Environment*, 4(3):144-53.

Welford, R. 1992. Linking quality and the environment: a strategy for the implementation of environmental management systems. *Business Strategy and the Environment*, 1(1): 25-34.

Welford, R. 1993. Breaking the link between quality and the environment: auditing for sustainability and life cycle assessment. *Business Strategy and the Environment*, 2(4): 25-33.

Welford, R. 1995. *Environmental strategy and sustainable development: The corporate challenge of the 21st century.* London: Routledge.

Welford, R. (Ed.). 1996. *Corporate Environmental Management: Systems And Strategies.* London: Earthscan.

Welford, R. & Starkey, R. (Eds.). 1996. *Business and the Environment.* London: Earthscan.

Winter, G. 1988. *Business and the environment.* Hamburg: McGraw-Hill.

Winter, M. 1997. *Oekologisches organizationslernen.* Wiesbaden: Gabler Verlag.

Winter, M. & Steger, U. 1997. Environmentally-motivated organizational learning. Unpublished working paper. Lausanne, Switzerland: *IMD*.

Winter, M. & Steger, U. 1998. *Managing outside pressure: Strategies for preventing corporate disasters.* West Sussex, England: Wiley & Sons.

Witt, L. A. & Beorkrem, M. N. 1989. Climate for creative productivity as a predictor of research usefulness and organizational effectiveness in R & D organizations. *Creativity Research Journal*, 2: 30-46.

Woodman, R. W., Sawyer, J. E. & Griffin, R. W. 1993. Towards a theory of organizational creativity. *Academy of Management Review*, 18: 293-321.

World Business Council for Sustainable Development. 1996a. *Eco-efficiency and cleaner production: Charting the course to sustainability.* Geneva, Switzerland: UNEP.

World Business Council for Sustainable Development. 1996b. *Eco-efficient leadership for improved economic and environmental performance.* Geneva, Switzerland.

World Business Council for Sustainable Development. 1996c. *Annual review.* Geneva, Switzerland.

World Commission on Environment and Development. 1987. *Our common future.* Oxford: Oxford University Press.

Xenergy 1997. *Green Energy Study.* Burlington, MA: Xenergy.

Ytterhus, B. E. 1997. The Greening of industry with a focus on eco-efficiency: the concept, a case and some evidence. Paper for the *13th EGOS Colloquium*. Budapest, July 3-5.

Yukl, G. A. 1986. *Leadership in organizations.* Englewood Cliffs, NJ: Prentice-Hall.

Zeffane, R., Polonsky, M. & Metley, P. 1995. Corporate environmental commitment: developing the operational concept. *Business Strategy and the Environment*, 3(4): 17-28.

Zuboff, S. 1988. *In the age of the smart machine.* Oxford: Heinemann Professional Publishing.

Notes

CHAPTER 1

1 Sustainable development is defined as ensuring that we meet the needs of the present without compromising the ability of future generations to meet their own needs (World Commission on Environment and Development, 1987). No firm is yet sustainable, but environmentally proactive firms have attempted to prevent pollution, to minimize resource use, and to redesign products and services so they have less impact on the natural environment, in order to move toward sustainable operations. Currently the movement toward sustainability is a process rather than a discrete end.

2 Please note that eco-innovations are creative ideas applied to solve a particular problem, which fits with the definition we are using for innovations. As such they are creative ideas (innovations) even if the original idea was borrowed from a different use.

3 We have been asked the question "if companies engage in eco-innovations because of environmental pressures, assuming that those pressures are the same for all companies, why are certain companies more proactive than others?" First, companies come under different pressures, as we will explain later in this chapter, so we can not assume that the pressures on companies are the same. Second, our research question does not involve the identification of affects of external pressures on company eco-innovation. Rather, we have focused on a different question concerning the effect of organizational encouragement on individual employee eco-innovations. Third, we have clearly stated that the scope of our research only involves environmentally-proactive companies. Those firms who have not reacted to external pressures by trying to achieve a policy of sustainable development fall outside the parameters of our study.

4 The point has been made that technological and social innovations may not follow the same patterns as each other. We agree that this may be true, but we have not explored the differences between these in the general innovation literature, as we did not think it was of central concern to our research question. From our reading of the literature, the same factors of organizational and supervisory support reoccur regardless of the type of innovation.

5 Generally firms in the developed world all come under some external pressure to protect the natural environment. The implications for firms not having an

environmental policy of environmental protection usually involve regulatory fines and poor publicity from concerned stakeholders.

6   Other sources addressing these issues include the book *Beyond the Limits* (1992) and the Group of Lisbon's (1995) book *Limits to Competition*. These emphasized the resource limitations and pollution problems originally brought to light by the Club of Rome.

7   We have presented the relevant literature describing the emergence of a movement that placed external pressures on firms to protect the natural environment in order to present a historical context for companies taking responsibility for pollution coming from their operations. We do not have an opinion about this movement, nor is the movement a central concern to our research question. Indeed a discussion of the above literature would be an interesting separate dissertation topic.

8   Environmental performance is the company's success, or lack there of, in protecting the natural environment from negative impacts, such as pollution or natural resources depletion, during its business operations. The environmental performance of a firm can extend to the total life cycle of its products and services. Environmental performance improves, for instance, if a company's products use less energy during use and are easily recyclable after use.

9   While it would be interesting to explore further the differences between regulations and recommendations by international bodies, we have focused our research in this section of the thesis on general trends in regulation that have influenced companies to become proactive. Also, the issue of non-compliance with regulations and sanctions would be of interest to explore, but since this is tangential to our focus we have not done so. Please note that we are concerned with proactive companies, who by definition have gone beyond compliance with regulations and have developed policies of sustainability. These companies all intend to comply with existing regulations. (See Hoffman 1997, for further discussion of regulatory enforcement beginning in 1960.)

10 Even companies not associated with a natural disaster began to be concerned about what would happen to their credibility, or "license to operate", should they have a highly publicized accident.

11 Using the typology of Fischer and Schot, we have included a description of environmental activist groups' role in eroding business credibility under our description of credibility pressures. It has been noted that one approach would have been to use a different typology, which separates the discussion of pressure groups from the ecological disasters that these groups helped to publicize. But, the author feels that these two aspects to credibility pressures (i.e. the disasters and the groups that help to make the public aware of these

disasters) fit well together. We have therefore decided to group both discussions under the heading of credibility pressures.

12 Many environmental groups also use lobbying and political pressures to force companies to protect the environment, but here we are concerned with focusing on those highly visible actions that drew the public's attention to companies' poor environmental performance.

13 For example, Body Shop makes it clear in marketing of some products that buying that product helps to save the Rainforests.

14 It has been noted that eco-marketing may just be a marketing "trick" in some cases. But the author believes there is irrefutable evidence that the companies noted (Electrolux, Interface, Neste, Patagonia, Philips, and Xerox) have all made significant improvements in the environmental performance of their products and services as a result of market pressures or opportunities.

15 Superfund is an environmental law in the United States, which makes companies liable for costs to clean up hazardous waste sites. (See Hoffman, 1997 for more details.) Similar legislation is being explored by the European Union.

16 Pressures led some proactive firms to support policies of sustainability and eco-innovation, as discussed in detail below.

17 For instance, the group has done a great deal of academic and practical work on the issue of eco-efficiency, a tool for finding environmental solutions that combines resource efficiency with new product design and production.

18 The above sections on the environmental movement and pressures on firms to protect the environment are not central to the research question of this thesis but have been provided at the request of the thesis committee to provide background on why companies began to eco-innovate.

19 No similar study was done in European companies, so we have noted this study of United States companies expecting that it may parallel the case of European companies.

20 This is mentioned here instead of above because it relates directly to firm actions, such as audits, performance criteria, managerial evaluation based upon environmental actions, which is the topic of this section. In previous sections we have treated issues as they related to general trends and general pressures.

21 The reason that some firms have decided to go beyond reactive approaches is open to supposition and could be the topic of an entirely different thesis. The author believes, based upon observations of companies but not upon academic

research, that most companies have become proactive because of chief executive leadership. (See Post and Altman (1996) for support for this argument.) We think that corporate leaders adopted policies of sustainability either if they believed that protecting the environmental was the "right thing to do", or if they believed that external pressures would only increase on their industry and they therefore wanted to be a market leader in environmental protection and product development.

22 In Chapter 2 we provide a discussion of sustainability policies, defining these policies based upon an extensive literature review. Policies that fit into this category include purchasing policy, fossil fuel use reduction policy, toxic chemical use reduction policy, etc.

23 These authors are demonstrating uses of eco-innovative technologies in actions that improve ecological sustainability.

24 Note that the title of this section is "eco-innovation as a means to achieving sustainability".

25 The work of the Body Shop with homeless, unemployed, and indigenous peoples is intended to address the environmental equity issue. The theory is that redistributing wealth to those who have not previously had a stake in the economic benefits of natural resource use can have a positive impact on the environment as these people will be less likely to destroy the natural environmental which is supporting them economically. For instance, rain forests will have a real economic value from employment for indigenous people, leading those people and the governments in those countries to respect the forests as a source of natural products, instead of deforesting the land to create farmland.

26 One possible reason that this group of companies is relatively small may be due to differing strategies, resulting from different perceptions about the natural environment. Companies with a green champion in top management or who perceive increasing regulatory and stakeholder pressures may be earlier to adopt these sustainability policies than other companies. Please note that it is not a central concern of our research to determine why companies have become proactive, but rather to test factors of support in this type of company. We have observed that companies who have policies of sustainability perceive that they need eco-innovation to achieve this policy.

27 Resources did not allow for a separate exploration of these two topics. We agree that a separate study of work challenge and work group supports would be of interest for future research, but to the author it is not immediately obvious how these two categories might have special significance in the environmental case.

28 As also stated in Taylor (1947) and Zuboff (1988).

29 Every employee, whether at a low or middle level in a company, has the power to influence some aspect of the business. This is an important assumption of the empowerment literature. In practice, the author believes this to be true. Specifically, an employee at any level can propose and implement an environmental solution, no matter if the employee is a blue-collar worker on the shop floor, or a secretary in an office.

30 Below we look specifically at factors that influence innovation, and we find that they parallel those factors that affect employee involvement.

31 See Vroom (1964) for more on expectancy theory.

32 Harrison (1983), Kanter (1983), and Walton (1985) all indicate that empowerment leads to increased employee motivation which in turn encourages commitment, risk-taking and innovation. The author has found no disagreement in the literature on this point. The extensive empowerment literature consistently points to the fact those employees, who are empowered, respond by being more likely to innovate. We can not think of situations where empowerment would not encourage employees to find innovative solutions to problems that face firms.

33 Van de Ven (1986) describes lack of leadership as a problem in organizations which want to encourage innovations. Specifically, leadership commitment and support for the innovation process are "critical for creating a cultural context that fosters innovation" (p. 601). Kelley (1976) affirms this conclusion.

34 Experimentation with alternative approaches is an important factor in finding innovative solutions to problems in a business setting. One way experimentation takes place in companies is that supervisors provide discretionary funds to an employee or team of employees to implement a new process or develop a new product idea on a small scale. Supervisory behaviors that support experimentation are considered helpful to innovation.

35 Even in flat organizations with limited hierarchy we have found that employees have someone who they consider to be their boss. This person is usually responsible for setting employee goals and responsibilities, evaluating employee job performance and providing incentives for reaching specific objectives.

36 Supervisors are defined in our study as the manager that the employee identifies as his or her boss. As explained above, supervisors are responsible for job performance evaluation, setting of employee goals and responsibilities

and providing incentives (rewards and recognition) for employee achievement of objectives.

37 Note that this is a new set of literature. The authors are talking specifically about supervisory roles in encouraging employees, not empowerment, which can be operationalized by supervisors, but is discussed above as an organizational goal.

38 Redmond, Mumford and Teach (1993) are talking about the role of supervisors in setting priorities for employees when there are too many competing activities vying for limited resources of time and attention.

39 We do not measure supervisory values in our study. But, the literature indicates that it can affect employee motivation to innovate. We feel managerial values may provide a reasonable explanation for why some supervisors in the same organization give differing levels of support to employee eco-innovation, despite managing in the same organizational context, with the same incentives, and environmental policies.

40 Employees who care about protecting the natural environment can influence firms to not pollute or to take a more proactive approach to managing the environmental impacts from business operations. It is not surprising that employees can be vocal environmental stakeholders as they usually live in the community surrounding the business operations and have a stake in reducing the environmental impacts on that community. They also have a stake in protecting themselves from exposure to hazardous materials in the work place.

41 Also see part one of this chapter where we discuss how stakeholder pressures affected corporate willingness to protect the environment. We cite the cases of environmental group, insurance company and financial institution pressures on companies.

42 We have been asked why the motivation for managing environment should affect how much is done or should affect the manner in which it is done. Rands (1990) supports our contention that people who do not value the natural environment are less likely to manage environmental protection activities. From personal experience, the author has observed that managers or employees who care about the natural environment are those who are most likely to make an effort to manage the companies' environmental impacts. In addition, March and Simon (1958) and Hostager et al (1998) discuss the affect of intrinsic motivation on employee actions. March and Simon note in the general management case that intrinsic motivation (described as the alignment of the job with the person's values) affects worker motivation to perform a job. Hostager et al note that intrinsic motivation (defined as values and capacity) affects employee eco-innovation.

43 Note that we do not include values in our empirical study, but since they are a variable that could affect supervisor behaviors and employee eco-innovation we have included a discussion of the literature here.

44 Our research looks at the factors that are controlled by the organization, but does not address the individual's existing capacity and intrinsic motivation (values), both of which are factors that can be affected by organizational support, but which might also exist regardless of organizational actions. To investigate the impact of employee values on eco-initiatives, while an interesting question for further research, is outside the scope of our research because it would require a different methodology and, possibly, different tools (such as in-depth interviews with employees).

45 These behaviors, as the literature on leadership styles notes, are important to support any type of corporate performance.

46 Please note that while the two sets of literature repeat the same factors, that is precisely what we wanted to demonstrate in part three of this chapter. Thus, any repetition of factors is not only necessary, but also extremely germane to the purpose of this literature review.

47 Amabile et al (1996) made this point for employee creativity, but Hostager et al and Keogh and Polonsky make the point specifically in the case of supporting employee eco-innovation.

48 These factors are those that we have seen affect general innovation and that the environmental literature notes are important for environmental change toward sustainability. This literature is unique in that it suggests that these factors will also be important for supporting employee participation in eco-innovations.

49 The work of Hostager et al (and other researchers in this section) is specifically directed at employee eco-innovation, and thus, this is the first time the noted set of factors has been directly linked to employee eco-innovation. Milliman and Clair mentioned training, performance appraisals that make employees accountable, and rewards that provide incentives. Storen states that information systems, performance measures and inspiration affect employee eco-innovation. Lefebvre et al state that integration into strategy and involvement of managers at all levels is important to employee eco-innovation. All of these are new points that have not previously been made in the thesis since the previous literature was not talking specifically about employee eco-innovation.

50 In general, management support, available resources, organizational structures, and incentives can all have a positive influence on employee willingness to participate in intrapreneurial behavior (i.e. behaviors from which new ideas and their implementation result) (Brazeal, 1993, 1996; Fry, 1987; Kanter,

1989b; Kuratko, Montagno, and Hornsby, 1990; Peters and Waterman, 1982; Pinchot, 1985; Pinchot and Pinchot, 1994; Schuler, 1986; Tushman and Moore, 1988; Van de Ven, Angle, and Poole, 1989).

51   This point was made with regards to values generally affecting environmental involvement of employees and managers, but here we are making the point that intrinsic motivation specifically affects employee willingness to eco-innovate (the subject of our empirical research).

CHAPTER 2

1   All of the companies in the sample had a written environmental policy. This written policy and / or other separate policies included most or all of the thirteen policies. We test employee perceptions of these policies in two ways. First we test for knowledge of the policies existence.   Second, we test employee perceptions of organizational commitment to the policies. We use a five-point scale to measure these two components of employee perceptions. Please refer to the methodology chapter where we describe the measures we use in detail and how we interpret these measures.

2   Note that other aspects of organizational encouragement, as discussed by Amabile, Conti, Coon, Lazenby and Herron (1996), are included in hypotheses 2-8 which tests for supervisory encouragement of employee creativity and innovation.

3   We do not approach the analysis with a preconceived notion that one policy will have a greater influence on employee eco-innovation than another will. But, since most of the policies may exist in the overarching written policy we think that policy may have a greater influence.   That said, we do place greater weight on any policy, but let the statistical analysis determine which policies influence employee actions.

4   We make this comparison using a chi-square test of difference.   For more details on our analytical tools used and approaches to verifying our results please refer to the chapter on methodology.

5   Note that while there is general agreement by environmental researchers, based upon case study evidence, that environmental policies can demonstrate organizational commitment, we made an empirical test of this supposition. The author believes that an important part of scientific discovery is testing the factual basis for conclusions assumed to be generally true, but that have not been rigorously tested.   In the case of environmental policies of sustainability, we find that even though researchers assumed they were important to support employee participation in environmental initiatives, our results show that most of these policies do not have a statistically significant affect.

6 Life cycle assessment or analysis (LCA) is a process for analyzing every stage and every significant environmental impact of a product from the extraction and use of raw materials through to the eventual disposal of the components of the product and their decomposition back to the elements (Welford, 1996; p. 140).

7 Sunderland (1996) notes that as of the date of his study, the EMAS regulation had been more widely adopted in Europe than ISO 14001, but according to Steger (1998b) ISO 14001 is becoming the market leader. Steger notes that EMAS has "significant benefits" and "could contribute towards more sustainable development" (p. 7).

8 At this point in time most companies are using ISO 14001 and therefore, since international companies are mostly choosing the same management system it is not important to analyze differences between the European and international environmental management systems. Companies in the United States have tended to adopt ISO 14001 since no national system existed.

9 Note that environmental management systems relate to sites or organizations, whereas life cycle analysis relates to products and process variables.

10 Note that providing information on environmental performance in an annual financial statement or report is different from publishing an annual or biannual report devoted entirely to the company's environmental performance.

11 Environmental managers with whom we have spoken confirmed that a written environmental policy and management systems were both effective means of implementing the other listed policies. This is not a disputed point.

12 Note that included in both the international charter and EMS / LCA lists (but not mentioned above) are policies on risk reduction and emergency preparedness. We considered both outside the purview of our research as neither policy influences employee eco-innovations.

13 In all cases we examined, companies had waste reduction targets in their environmental performance goals since this is one sure way for companies to save money, by saving resources. Thus companies are highly motivated to reduce waste as a first step in any environmental policy application.

14 As mentioned above, the specific content of the written environmental policy differed from company to company, but we only included companies in our survey whose policy was to go beyond compliance toward sustainable development.

15 Targets are used to continuously improve company environmental performance. Each company is in a unique position to assess the best stretch

targets for themselves. Since only proactive companies were in the survey, we did not try to assess the quality of each company's targets.

16 An example of a service with reduced impact might be using rail transportation to deliver a product to market instead of trucks.

17 There is a definitional difference between employee responsibility and involvement. The notions of obligation and accountability are attached to responsibility. Involvement is the engagement of the interests or commitment of one. Therefore an employee could be involved without feeling responsible or accountable for an outcome.

18 We have developed the actual behaviors in a BARS development process, which we describe in detail in the methodology chapter (3). Therefore, we have placed the list of rank ordered BARS in that chapter as Table 3.1.

19 In our case, we are interested in environmental values.

20 Included in the category of communication are behaviors that demonstrate openness to employee communication, and behaviors that reinforce or break down organizational hierarchy, which are related to supervisory reinforcement of open communication.

21 We tested the assumption made in the literature that the behaviors that support general activities could also be those that support environmental activities. Since the BARS are general in nature, it is possible for us to ask two questions to test this assumption, one to find out from respondents the behavior their supervisor uses in general, and another to find the behavior used in the environmental case. The same parallel between general organizational policies and environmental policies does not exist. There is no reason for us to believe that strong human resource policies, for instance, would necessarily signal organizational support for environmental initiatives by employees. Therefore, we did not attempt to make any comparison between general organizational policies (on any subjects unrelated to environment) and environmental policies.

CHAPTER 3

1 Please note that all but one of the companies in our sample have between 3200 and 41,000 employees. The company with 1500 employees hires approximately 2300 contractors (from a diversity of nationalities) on average per year, and holds these contract employees to the same high level of environmental performance as their permanent employees. For instance, they hold environmental training courses for both contract and permanent employees.

2   We feel it is reasonable to include employees from diverse work force units and functions in our sample because we are interested in testing if policies and behaviors impact employee eco-innovations of different employees throughout different companies. Note that the environmental policies are the same for the entire company and that the behaviors are general in nature. Also note that environmental initiatives, while they may differ from one location in the company to another, are possible in all functions of a company. We are not arguing that the same type of eco-innovation would take place in accounting as in production, but neither are we concerned with the specific types of eco-innovations that are occurring.

3   We did not ask respondents to self-indicate demographic information because we believed that respondents would feel that not self-identifying increased the confidentiality of their responses. With a survey of this type, where employees are asked for candid assessments of their direct supervisor's behavior, we felt that it was important that employees feel that the questionnaires could not be linked back to them. We can not think of any case where information on age or sex, for instance, would have significant bearing on our research question.

4   We needed a sample where all of the companies stated that they were committed to sustainable development, and most of which had all thirteen policies we identified as possibly signaling organizational commitment to environmental innovation. Including companies without these proactive policies would have served no useful purpose. Indeed, our sample allows us to test our research question because it includes companies with these policies.

5   Interestingly we are able from the results to draw some conclusions not only about why employees eco-innovate, but also about why employee are not eco-innovative.

6   These BARS were later placed on a ten point scale for the analysis.

7   Note that since this was confidential information, we can not reveal it in a published document, such as this dissertation.

8   Eco-initiative is a term readily understood by employees in European companies in general. And, there is a history of formalized proposal systems (such as suggestion boxes) in all of the companies in the sample. Therefore, we feel this term led to easy and understandable interpretation of the dependent variable question by respondents in the survey.

CHAPTER 4

1   This relationship is more complex than a simple correlation as it is a non-linear relationship.

2   Note we recognize that including these screening questions in the survey may have deterred some employees from responding to the survey questionnaire. If so, those employees who chose not to respond may have tended to be employees with negative answers to the screening questions. As such, the total sample may be biased toward employees with positive opinions of general management and of the companies economic health. But, since we found no statistically significant difference between the responses of those who responded negatively to the screening questions and the rest of the employee responses, we do not feel this created a bias in the overall sample.

3   Note that even though we selected six environmentally-proactive companies, all of which stated in interviews that they had a strong policy commitment to sustainable development, we found that not all had sustainable development policies of the types described in the literature. (We had assumed that not all of the companies would have all of the policies we selected, but all companies had most of the policies and some had all of the thirteen policies.) As noted above in the Methodology chapter, all six companies had policies for questions 1-7 and 13; one company did not have a policy for 8 (use of Life Cycle Analysis); 9 (management understands sustainable development and is addressing it); and 11 (toxic chemical use reduction); three companies did not have a policy for 12 (reducing use of unsustainable products); and, four companies did not have a policy for 10 (fossil fuel use reduction). Due to the absence of some of the sustainable development policies in some of the companies, we expected weaker statistical results for questions 8, 9, 10, 11, and 12. This is borne out in our analyses below.

4   An explanation of this exception is addressed below.

5   Alternatively, this exception could show a weakness in the survey questionnaire. The wording of the information dissemination behaviors may have been too general for employees to apply them to the environmental case.

6   Please note that the literature related to perceptions indicates that people often perceive what they want to perceive. Therefore, though we found a systematic and statistically significant result, individual results would be expected to vary.

CHAPTER 5

1   Note that while our research was focused on the case of environmental innovation, we believe that the behaviors in the BARS survey are effective at testing for the general case. Therefore we have used these results to draw some general conclusions about learning in organizations as shown above.

Appendix

Behavioral Survey Development

In the Appendix we have included materials that supported the development of the behaviorally-anchored rating scales. This process included interviews with employees at companies with excellent environmental practices, an allocating exercise performed by environmental managers from European corporations, a scaling exercise performed by employees at two European-based companies, and the testing and finalization of a questionnaire. This final questionnaire measures employee perceptions of organizational support for environmental management and supervisory support for environmental activities by employees. A detailed description of the survey development process can be found in Chapter 3.

Appendix 1.1: Interview Questionnaire

Appendix 1.2: Instrument for Allocating Performance Characteristics to Managerial Behaviors

Appendix 1.3: Scaling Exercise Questionnaire

Appendix 1.4: Final Questionnaire: Employee Assessment of Environmental Management Behavior

Appendix 1.1

Interview Questionnaire

To Facilitate the Development of a Behavioral Measurement Tool

This is a framework for a one hour interview. The interview is intended to discover positive (and less than positive) behaviors displayed by management teams in the following functional areas: Communication, Information, Innovation & Learning, Rewards & Recognition, and Measurement.

Discussion of Management Structure

First, we would like to find out about your "management team". This background information will provide a framework in which to understand your descriptions of behaviors and your other responses.

1. What are your responsibilities at work?

2. How many managers do you work with here at this facility?

Communication

We are interested in this organization's approaches to communication. For instance, we would be interested to know if communication is open and welcome, even about unpopular subjects. Are managers in your team open to criticism and new ideas? Do your managers encourage both positive and not so positive feedback? Do your managers trust you to make good decisions and empower you to make independent choices about how to do your job?

1. If you were to approach a manager with a new idea about how to improve your company's efficiency, what would be the likely response?

2. If you were to approach a manager with an idea of how to improve your efficiency, but it might cost the company money to implement your idea, what would be the likely response?

3. If you were to approach a manager with an idea about how to make an improvement in a functional area completely unrelated to your own, what would be the likely response? How would the manager show support (or no support) for your initiative? Would you be encouraged or discouraged from talking to the group for which you had the suggestion? Would it be likely that the manager would want to take the idea to the group him/herself?

4. How does your management team encourage dialogues about topics? Do you feel encouraged to discuss topics, even if you disagree on the probable outcomes?

5. If you had a criticism about a manager's work behavior, or a suggestion about how (s)he could improve his/her performance, what would be the most likely response to your telling the manager what you think?

6. If you have failed to meet an important deadline for the company, what would be your managing team's response?

7. If you have something important to tell to a manager on your management team, would you feel that the manager would listen attentively? Would you be confident that you would have access to the manager with whom you needed to talk?

8. Can you describe some defensive behaviors which you have observed from people within your management team?

Information

We are interested in knowing about the information systems that exist in this organization. Specifically, about how easy it is to get information from different sources and about the transparency of that information.

1. In general, if you wanted to find out something unrelated to your direct responsibilities, would you know where to go to look for the information? How would people in the company react to your desire for that information? For example, if you wanted to know the environmental emissions of a sister facility, would you have easy access to that information?

2. If you wanted to go to an area within your facility or to a competing facility to try to learn how they perform a function similar to your own, how would your management team respond?

3. If you asked someone on your management team about what had occurred in a recent meeting to which you were not invited, what would be the manager's probable response?

4. If you wanted an appointment with the CEO of your company, how would your management team react? How would the CEO's office react? How would the CEO him/herself react to your visit (should it be possible)?

5. If there is a change in company policy, how would that information be communicated? What would be the approach and attitude of the manager relaying the information? What would the managers response be to delving, inquisitive questions about the new policy?

6. Do you ever feel that you are given too much information? For example, too many new directives and too much to remember? How would your management team react, if you told them that you were "drowning in" information? Is there any attempt to "unlearn within your organization?

7. Is information about your plant freely available to your community? How does management react to requests for information from outside your facility?

Learning/Innovation

In the area of Learning and Innovation we are interested in finding out how new ideas are treated within your organization. For instance, are they nurtured and how? Also, how information is shared to enable others in the organization to learn?

1. How does your management team assess your learning needs?

2. How is continuing education encouraged or discouraged by your managers' behaviors? How do your managers encourage (or discourage) you from getting knowledge from outside the organization?

3. Do employees learn to work in teams? When an employee offers an unpopular suggestion during a meeting, how does the group react?

4. How does your management team encourage you to work in teams?

5. How does your management react to changes, i.e. are they open to new ways of doing things? Which behaviors of your management team show that they are adaptable to changes and new ideas? Which behaviors of your management team show that they are slow to change or adapt to new situations and or knowledge?

6. Are you encouraged by your management team to rotate into new assignments? If you were to suggest such a change, how would your manager react?

7. If you were to suggest that a group experiment with a new way of doing things or a process change in order to improve quality or environmental performance, how would your management team react to your suggestion? How would the other employees effected by the change react?

8. Which behaviors displayed by your management team, if any, encourage risk-taking in your job with regards to new ideas or changing processes?

9. How does your management team react when you make a mistake?

Rewards and Recognition

We are interested in finding out how your management team rewards and recognizes innovation on the part of its employees.

1. If you were to create a more efficient, cost-saving way of doing business, how would your management team recognize your efforts?

2. How, in general, does your company reward innovations? How would managers decide which kinds of innovative performance to recognize?

3. Do your managers have informal behaviors which recognize positive achievements? Formal behaviors?

4. Has you company linked innovative performance to pay?

5. What are some of the ways your management team monitors your performance?

Measurement

We are interested in how your organization sets goals for its employees.

1. Do you feel you have clear and measurable responsibilities in your daily work? How does your management team make your responsibilities clear and measurable for you?

2. How does your management team measure your performance? Do they use qualitative measures and/or quantitative measures?

3. What types of goals are set for you, if any?

4. How regularly would you meet with someone from your management team to either formally, or informally, review your performance goals? What types of behaviors reinforce your achievement of your measurable goals? What types of behaviors discourage you from achieving your goals?

5. Which approaches do your managers use to set new goals for you? How would this goal setting be influenced by policies with in the company? Do all managers set measurable goals for their employees within your company? If not, how do those managers differ for those who do set measurable goals?

Appendix 1.2

Instrument for Allocating Performance Characteristics to Managerial Behaviors

Your goal is to allocate characteristics of managerial performance to behavioral statements which best describe them. The performance characteristics define some qualities of managers' styles which support or fail to support employee participation and initiatives in companies. See the following page for the entire list performance characteristics (examples: communication skills, use of rewards & recognition).

This exercise will take you 25-30 minutes and will be an important input into the development of the MIBE Project's Behavioral Survey. The results of your allocations will be the basis of the questions on the survey. Specifically, under each performance characteristic there will be several good, average and poor behaviors from which employees can choose the example which most closely describes their management team.

For this allocation exercise, when pairing the behaviors with the performance characteristics you will find examples of supportive and unsupportive behaviors. There are no right or wrong answers. Your objective is to match these behaviors with the most specific performance characteristic possible. In some cases you will see that the behaviors are specific enough for you to assign both **a letter (A-E) and a number (1-26)**. In other cases, you will find that the behavior describes the entire set of performances and you will only assign a letter to that behavior. Always be as specific as possible, i.e. when possible, match the behaviors with a letter and number. There will be both negative and positive behaviors described.

To better illustrate this exercise, suppose the first behavior listed was:

_____ *1. Uses bonus system to reward employees who have achieved their goals.*

If you wrote "D/21" in the space to the left of the example, this would mean that you believe that this behavior best represents or describes the performance characteristic of "Rewards and Recognition: offers financial rewards and pay for performance". **Note:** A "D" alone would also have been an appropriate response.

Please read through the performance characteristics on the following page. Note that there are quite a few more descriptors under some performance characteristics then others. Then go through the list of 128 behaviors which follows the performance characteristics, and allocate one of the characteristics to each of the listed behaviors by using a letter and/or number. Assign the specific performance characteristic which you think is best described by the behavior. Please do not skip any behaviors.

DESCRIPTION OF MANAGERS' PERFORMANCE CHARACTERISTICS
(Describes "What" Managers Do In Learning Organizations)

A. Innovation and Competence Building

1. Uses multi-level and cross-functional teams
2. Makes it acceptable to take risks
3. Focuses on the customer rather than the bottom line
4. Implements employees' suggestions
5. Sees mistakes as learning opportunities
6. Uses training and education to foster innovation
7. Fosters experimentation to develop new processes
8. Learns from outside the organization
9. Uses assignment rotations as learning tools
10. Quickly implements changes

B. Communication

11. Encourages individuals and groups of employees to communicate problems and propose solutions
12. Develops open and direct channels for communication
13. Engages in open-ended dialogues about strategic business issues
14. Holds discussions about failures and mistakes with employees
15. Practices listening skills
16. Facilitates communication between groups at all hierarchical levels
17. Supports criticism and dissent, both of self and of ideas in general

C. Information Dissemination

18. Provides clear, accurate, easily accessible information to internal/external audiences
19. Manages information flow by establishing what employees need to know
20. Informs employees where to find out what they want to know

D. Rewards and Recognition

21. Offers financial rewards, like pay and bonuses linked to performance targets and corporate goals
22. Uses awards
23. Uses informal feedback and acknowledgment of good ideas/actions

E. General Management/Measurement

24. Establishes clear responsibilities
25. Defines measurable goals, which include quantitative and qualitative dimensions
26. Provides feedback in order to help employees achieve goals

LIST OF MANAGERS' BEHAVIORS
(Describes "How" Managers Do Things)

_____ 1.   Creates an open environment in which to discuss decisions which affect the business.

_____ 2.   Often keeps the door to his/her office closed.

_____ 3.   Supports a long and precise planning process.

_____ 4.   Listens to employees, then forgets/ignores what they have said.

_____ 5.   Encourages employees to express concerns about company decisions and policies.

_____ 6.   Keeps responsibility for all decisions.

_____ 7.   Acts suspicious of other areas of the company.

_____ 8.   Meets regularly with employees to discuss progress toward their individual goals.

_____ 9.   Trusts employees to do a professional job and respects skills of employees.

_____ 10. Encourages employees to spend time in training and education courses, including providing funding for work related courses.

_____ 11. Resists new ideas, independent of the source of the proposed change.

_____ 12. Avoids difficult decisions by postponing discussion.

_____ 13. Commits to an action, but then does nothing.

_____ 14. Reinforces organizational hierarchies by insisting that employees be of the same level in order to communicate.

_____ 15. Acts jealous when another division does well within the company.

_____ 16. Refuses to commit resources and employee time for training and education activities.

_____ 17. Rewards a good idea by implementing it and giving responsible employee credit.

_____ 18. Seldom, if ever, rewards an employee for work well done.

_____ 19. Often reschedules employee meetings (or cuts them short) in order to meet with other managers of the same level or higher than him/herself.

_____ 20. Discourages team work by keeping employees too busy to meet regularly.

_____ 21. Sends employees to other locations in the company and elsewhere to learn about innovative processes and other ways of doing business.

_____ 22. Finds opportunities to work alongside employees in order to support and monitor progress toward employees' goals.

_____ 23. Candidly discusses unacceptable behaviors with employees and, when needed, gives specific feedback on expected improvements.

_____ 24. Frequently asks for suggestions from employees.

_____ 25. Seldom offers to help, even if an employee is under pressure from a deadline.

_____ 26. Regularly asks employees about their ideas for training and education which the employees feel they want/need in order to perform their job better.

_____ 27. Asks specific questions about progress regularly in order to encourage a dialogue about the employee's work.

_____ 28. Sometimes lies to employees and/or others.

_____ 29. Looks for opportunities to praise employee performance.

____ 30. Uses company award systems to recognize particularly good performance of employees.

____ 31. Challenges underlying assumptions when discussing most topics.

____ 32. Won't point out unacceptable behavior to employees for fear of being unpopular.

____ 33. Often criticizes or finds fault with others.

____ 34. Seldom admits to a mistake s/he has made.

____ 35. Blames others when s/he makes a mistake or fails to perform well.

____ 36. Works with employees as an equal member of the team.

____ 37. Clearly has favorite employees and treats them better than others.

____ 38. Delegates specific tasks to employees and tells them precisely how the tasks should be performed.

____ 39. Discusses with clarity, both short- and long-term organizational and individual employee goals.

____ 40. Tries to hide unpopular decisions and information from employees.

____ 41. Is often vague about what s/he wants from an employee.

____ 42. Sometimes criticizes an employee in front of others.

____ 43. Frequently delays or avoids discussions of mistakes as well as other discussions where conflict might occur.

____ 44. Frequently keeps door open.

____ 45. Seldom provides specific feedback on employee performance.

_____ 46. Avoids discussing poor employee performance until the annual review, or after it is too late for the employee to achieve a promotion or a pay increase.

_____ 47. Would dismiss an employee who failed to meet an important company deadline.

_____ 48. Often interrupts someone who is trying to explain an idea, concern, or problem.

_____ 49. Frequently shows support for an employee's idea by communicating the suggestion to others, implementing or helping the employee to implement the idea.

_____ 50. Usually objects to changes and new ideas and finds excuses why they can not be implemented.

_____ 51. Usually thanks an employee for his/her suggestion, whether it can be adopted or not.

_____ 52. Encourages employee trust by openly announcing information without delay about troubling situations, like lay-offs or restructurings.

_____ 53. Clearly explains the reason for the organization's goals or policies and forewarns employees about expected changes whenever possible.

_____ 54. Seldom has time available for discussions.

_____ 55. When meeting with employee(s), allows frequent interruptions and shows lack of focused attention on what employee(s) are saying.

_____ 56. Demonstrates that topic and discussion are important by clarifying his/her understanding and showing respect toward the employee presenting the information.

_____ 57. Uses all available methods to spread the word about company information, including meetings, bulletin boards, computers, etc.

____ 58.　Seldom praises employees' work.

____ 59.　Normally penalizes employees for failure, either with negative judgments on performance evaluations or by yelling at them.

____ 60.　Regularly spends time discussing issues, challenges and opportunities with employees.

____ 61.　Tries to manage every detail of an employee's work so that the employee has little freedom to do his/her job independently.

____ 62.　When asked a direct question, often finds a way to avoid answering it directly and honestly.

____ 63.　Secretly checks on employees in order to find fault or to catch them doing something wrong.

____ 64.　Usually documents mistakes on the employee's record.

____ 65.　Announces and praises employees' positive contributions, including new ideas and cost-savings initiatives, to others.

____ 66.　Discourages employee trust by using job performance information to negatively affect someone else's opinion of that employee.

____ 67.　Encourages employees to solve problems using teams.

____ 68.　Gets vindictive and sometimes refuses to talk with an employee after the employee has offered a suggested improvement or criticism of the manager's job performance.

____ 69.　Spends time discussing and implementing a learning plan with each employee.

____ 70.　Encourages participation on cross-functional and/or multi-level teams. (I.e. teams with participants from other parts of the company and/or with both managers and employees working together.)

_____ 71. Frequently admits when s/he does not know something or when s/he has been wrong.

_____ 72. When deciding to make changes, tries to collect complete information and accurate data before making a decision.

_____ 73. Keeps information about problems in our area private from the rest of the company and tries to solve them without help.

_____ 74. Often experiments, on a small scale, with new ideas in order to examine if the possible change should be adopted on a larger scale.

_____ 75. Often takes credit for someone else's good idea.

_____ 76. Seldom reacts negatively to employees' or managers' questions and/or concerns.

_____ 77. Demonstrates caring and concern when interacting with employees.

_____ 78. Seldom experiments with new ideas or methods of doing things.

_____ 79. Often challenges employees to think and act in new ways.

_____ 80. Realigns employee responsibilities to allow employees time for training, site visits, or exploring new techniques for doing their job.

_____ 81. Seldom talks to employees about goals and responsibilities, except when required by company policies.

_____ 82. Quickly responds to employee ideas and suggestions, even if they are not adopted.

_____ 83. Rigidly adheres to deadlines.

_____ 84. Talks regularly with employees to assess progress toward explicit employee goals.

____ 85. Is offended if someone tells him/her how to do his/her job.

____ 86. Uses both quantitative (numbers) and qualitative (quality) measures to assure individual is making progress toward or contributing to company goals.

____ 87. Delays giving employees the training/education they need when there are changes in their job functions.

____ 88. Answers questions honestly even if the answer is not what the employee wants to hear.

____ 89. Does nothing to avoid or to manage employee information overloads.

____ 90. Uses information systems, such electronic bulletin boards, videos, computer systems, etc. to share learning amongst employees.

____ 91. If there is too much information, s/he sets priorities and establishes what is most important for employees to know.

____ 92. Uses dialogue with the community to help challenge the company to continuously improve.

____ 93. Involves employees in changes by instilling ownership of problems and responsibilities for solutions in every employee.

____ 94. Tries to control everything employees do and is afraid to allow employees to try new ways of doing their jobs.

____ 95. Makes quick decisions, but without gathering all the facts beforehand.

____ 96. If the company does well, management will reward all the employees.

____ 97. Listens openly and attentively to suggested improvements in how s/he does his/her job and often adopts the suggestions.

_____ 98. Listens to and values input from employees and managers from other parts of the company.

_____ 99. Tries to prevent participation from disruptive, questioning employee by not inviting him/her to meetings.

_____ 100. If a mistake was made, would transfer the responsible employee to another area in the company.

_____ 101. Uses individual and measurable goals as basis for encouraging employee responsibility in the workplace.

_____ 102. Uses bonus system to reward employees who have achieved their goals.

_____ 103. Encourages employees to search for information, even if it is unrelated to their job responsibilities.

_____ 104. Resents questions from employees.

_____ 105. Encourages partnerships with other departments in order to implement new ideas.

_____ 106. Listens to criticisms, but never changes behavior.

_____ 107. Finds help for employees, if they are having difficulty meeting a deadline.

_____ 108. Usually encourages participation in any learning situation in which an employee would like to engage.

_____ 109. Avoids difficult discussions and ignores problems as they are developing.

_____ 110. Penalizes employees for missed deadlines.

_____ 111. Would accompany an employee to discuss and promote the employee's idea with another manager.

_____ 112. Freely involves outside stakeholders in discussions about the company.

____ 113. Tells employee quickly when there is something wrong with his/her work.

____ 114. Forms a new group if s/he doesn't like the decision of the existing team.

____ 115. Often brings in people from other parts of the company when they have knowledge which can help our group solve a problem or improve our decision-making.

____ 116. I would never approach my management team with a suggested change because I know they would be angry with me for interfering in their area of responsibility.

____ 117. Management requires that we spend lots of time each year planning for the next year.

____ 118. If there is something new I need to know, management will make sure I have training/education on it quickly.

____ 119. I think that this company hides information from employees and the external community.

____ 120. If I make a mistake, I might get fired.

____ 121. I feel strong pressure not to approach any manager, other than my direct supervisor, even if I feel someone else needs to hear my idea or concern.

____ 122. If I show management a better way to do a job, they will encourage me and others to adopt the change.

____ 123. I need an appointment if I want to talk with a manager in this company.

____ 124. When someone makes a mistake, we usually discuss, as a group, how to avoid the problem in the future.

____ 125. Our group is often the last to know about changes in the company because our management team does not tell us things.

_____ 126. Management acts like they are more important than the employees.

_____ 127. If I was found talking to employees from a different area of the company, management would assume that we were not talking about work and might reprimand me.

_____ 128. If I wanted a meeting with the head of the company, my management would discourage me from making the appointment.

_____ 129. If I were to have a suggestion for how management could improve, management would be angry with me for telling them.

_____ 130. If I do a good job, I am not certain that management will notice. But, if I make a mistake, I am sure that management will notice and probably reprimand me.

_____ 131. Discourages me from asking questions about other parts of the company or about activities which are outside my areas of responsibility.

_____ 132. Frequently finds something good to say about how I am doing my job.

Thank you.

Scaling Exercise Questionnaire

This exercise is an input into the creation of a behavioral survey tool which will be administered at companies who wish to discover areas where they can improve their management support of employee involvement in company environmental programs.

Below is a rating scale to be used in order to rate the attached list of examples of managerial behaviors. Your goal is to place a numerical value, using the guidelines below, in front of each of the attached examples. This number represents your judgment as to the degree of managerial effectiveness demonstrated by the particular behavioral example. There are no right or wrong answers.

Here is an example to clarify these instructions:

Communication

Examples Related to Management Team's Support of Communication

_____ 1.   Creates an open environment in which to discuss decisions which effect the business.

If you placed a "4" in front of the above behavior, that would mean that you think that the behavior described is an example of "effective" managerial support for communication.

This exercise will take you approximately 20 minutes. Thank you for your participation.

Rating Scale - Degree of Managerial Support for Learning Functions Within Their Company

Scale Value	Definition of Scale Value
1	Ineffective
2	Low effectiveness
3	Moderately effective
4	Effective
5	Highly effective

A. Innovation

<u>Examples Related to Management Team's Support of Innovation</u>

_____ 1.   Resists new ideas, independent of the source of the proposed change.

_____ 2.   Sends employees to other locations in the company and elsewhere to learn about innovative processes and other ways of doing business.

_____ 3.   Frequently asks for suggestions from employees.

_____ 4.   Works with employees as an equal member of the team.

_____ 5.   Frequently shows support for an employee's idea by communicating the suggestion to others, implementing or helping the employee to implement the idea.

_____ 6.   Usually objects to changes and new ideas and finds excuses why they can not be implemented.

_____ 7.   Encourages employees to solve problems using teams.

_____ 8.   Encourages participation on cross-functional and/or multi-level teams. (I.e. teams with participants from other parts of the company and/or with both managers and employees working together).

_____ 9.   Often experiments, on a small scale, with new ideas in order to examine if the possible change should be adopted on a larger scale.

_____ 10.   Seldom experiments with new ideas or methods of doing things.

_____ 11.   Often challenges employees to think and act in new ways.

_____ 12.   Quickly responds to employee ideas and suggestions, even if they are not adopted.

_____ 13.   Encourages partnerships with other departments in order to implement new ideas.

_____ 14.   Would accompany an employee to discuss and promote the employee's idea with another manager.

_____ 15.   Often brings in people from other parts of the company when they have knowledge which can help our group solve a problem or improve our decision-making.

_____ 16.   I would never approach my management team with a suggested change because I know they would be angry with me for interfering in their area of responsibility.

_____ 17.   If I show management a better way to do a job, they will encourage me and others to adopt the change.

_____ 18.   When someone makes a mistake, we usually discuss, as a group, how to avoid the problem in the future.

B. Competence Building

<u>Examples Related to Management Team's Support of Learning</u>

_____ 1.   Encourages employees to spend time in training and education courses, including providing funding for work related courses.

_____ 2.   Refuses to commit resources and employee time for training and education activities.

_____ 3.   Regularly asks employees about their ideas for training and education which the employees feel they want/need in order to perform their job better.

_____ 4.   Spends time discussing and implementing a learning plan with each employee.

_____ 5.   Realigns employee responsibilities to allow employees time for training, site visits, or exploring new techniques for doing their job.

_____ 6.   Delays giving employees the training/education they need when there are changes in their job functions.

_____ 7.   Usually encourages participation in any learning situation in which an employee would like to engage.

_____ 8.   If there is something new I need to know, management will make sure I have training/education on it quickly.

C. Communication

Examples Related to Management Team's Support of Communication

_____ 1.   Creates an open environment in which to discuss decisions which effect the business.

_____ 2.   Often keeps the door to his/her office closed.

_____ 3.   Listens to employees, then forgets/ignores what they have said.

_____ 4.   Encourages employees to express concerns about company decisions and policies.

_____ 5.   Reinforces organizational hierarchies by insisting that employees be of the same level in order to communicate.

_____ 6.   Often reschedules employee meetings (or cuts them short) in order to meet with other managers of the same level or higher than him/herself.

_____ 7.   Frequently keeps door open.

_____ 8.   Often interrupts someone who is trying to explain an idea, concern, or problem.

_____ 9.   Seldom has time available for discussions.

_____ 10.   When meeting with employee(s), allows frequent interruptions and shows lack of focused attention on what employee(s) are saying.

____ 11.   Demonstrates that topic and discussion are important by
           clarifying his/her understanding and showing respect toward
           the employee presenting the information.

____ 12.   When asked a direct question, often finds a way to avoid
           answering it directly and honestly.

____ 13.   Seldom reacts negatively to employees' or managers' questions
           and/or concerns.

____ 14.   Answers questions honestly even if the answer is not what the
           employee wants to hear.

____ 15.   Listens openly and attentively to suggested improvements in
           how s/he does his/her job and often adopts the suggestions.

____ 16.   Listens to and values input from employees and managers from
           other parts of the company.

____ 17.   Resents questions from employees.

____ 18.   Avoids difficult discussions and ignores problems as they are
           developing.

____ 19.   I feel strong pressure not to approach any manager, other than
           my direct supervisor, even if I feel someone else needs to hear
           my idea or concern.

____ 20.   I need an appointment if I want to talk with a manager in this
           company.

____ 21.   If I wanted a meeting with the head of the company, my
           management would discourage me from making the
           appointment.

D. Information Dissemination

<u>Examples Related to Management Team's Use of Information Systems</u>

____ 1.   Encourages employee trust by openly announcing information without delay about troubling situations, like lay-offs or restructurings.

____ 2.   Uses all available methods to spread the word about company information, including meetings, bulletin boards, computers, etc.

____ 3.   Keeps information about problems in our area private from the rest of the company and tries to solve them without help.

____ 4.   Does nothing to avoid or to manage employee information overloads.

____ 5.   Uses information systems, such electronic bulletin boards, videos, computer systems, etc. to share learning amongst employees.

____ 6.   If there is too much information, s/he sets priorities and establishes what is most important for employees to know.

____ 7.   Tries to hide unpopular decisions and information from employees.

____ 8.   Clearly explains the reason for the organization's goals or policies and forewarns employees about expected changes whenever possible.

____ 9.   I think that this company hides information from employees and the external community.

____ 10.   Our group is often the last to know about changes in the company because our management team does not tell us things.

E. Rewards and Recognition

<u>Examples Related to Management Team's Use of Rewards and Recognition</u>

_____ 1.   Rewards a good idea by implementing it and giving responsible employee credit.

_____ 2.   Seldom, if ever, rewards an employee for work well done.

_____ 3.   Looks for opportunities to praise employee performance.

_____ 4.   Uses company award systems to recognize particularly good performance of employees.

_____ 5.   Seldom praises employees' work.

_____ 6.   Usually documents mistakes on the employee's record.

_____ 7.   Announces and praises employees' positive contributions, including new ideas and cost-savings initiatives, to others.

_____ 8.   Uses bonus system to reward employees who have achieved their goals.

_____ 9.   Often criticizes or finds fault with others.

_____ 10.   Penalizes employees for missing a deadline.

_____ 11.   Normally reprimands employees for failure, either with negative judgments on performance evaluations or by yelling at them.

_____ 12.   Frequently finds something good to say about how I am doing my job.

_____ 13.   If the company does well, management will reward all the employees.

_____ 14.   If I do a good job, I am not certain that management will notice. But, if I make a mistake, I am sure that management will notice and probably reprimand me.

F. Management of Goals and Responsibilities

Examples Related to Management Team's Monitoring and Setting of Goals and Responsibilities

_____ 1.   Tells an employee right away when there is something wrong with his/her work.

_____ 2.   Keeps responsibility for all decisions.

_____ 3.   Seldom offers to help, even if an employee is under pressure from a deadline.

_____ 4.   Delegates specific tasks to employees and tells them precisely how the tasks should be performed.

_____ 5.   Seldom provides specific feedback on employee performance.

_____ 6.   Avoids discussing poor employee performance until the annual review, or after it is too late for the employee to achieve a promotion or a pay increase.

_____ 7.   Seldom talks to employees about goals and responsibilities, except when required by company policies.

_____ 8.   Uses both quantitative (numbers) and qualitative (quality) measures to assure individual is making progress toward or contributing to company goals.

_____ 9.   Uses individual and measurable goals as basis for encouraging employee responsibility in the workplace.

_____ 10.   Is often vague about what s/he wants from an employee.

_____ 11.   Finds opportunities to work alongside employees in order to support and monitor progress toward employees' goals.

_____ 12.   Discusses with clarity, both short- and long-term employee goals.

_____ 13.   Tries to manage every detail of an employee's work so that the employee has little freedom to do his/her job independently.

_____ 14.   Talks regularly with employees to assess progress toward explicit employee goals.

_____ 15.   Involves employees in changes by instilling ownership of problems and responsibilities for solutions in every employee.

Final Questionnaire: Employee Assessment of Environmental Management Behavior

This questionnaire is **confidential**. Only the final results, not the names of the respondents, will be reported to your company. The purpose is to give your management information which will help them become more supportive of employee participation in the company's environmental program. This questionnaire will take approximately ten to fifteen minutes to complete.

PART I - GENERAL INFORMATION

1. Company name:_____

2. Division with which you work:_____

	Very Poor	Poor	Average	Good	Excellent
3. What is your general opinion of management in your company? (Circle a number.)	1	2	3	4	5
4. How do you feel about your future prospects in the company?	1	2	3	4	5

5. Has the company had major lay offs during the last three years?
   Yes _____ No _____

6. Has the company been profitable and/or growing over the last three years?
   Yes _____ No _____

PART II - INSTRUCTIONS

Company Support for Environmental Initiatives

The aim of this section is to measure your organization's commitment to environmental protection and employee involvement in the company's environmental program.

Questions 1 through 13 are general questions about your entire company's environmental policies and actions, whereas 14 and 15 are specific to your manager's support of your environmental initiatives.

If you respond to statements 1 to 13 in the order given on the next page, it may become increasingly difficult to answer 'agree' or 'strongly agree'.

Note that a choice of #3 means that you "don't know".

PART II - COMPANY ENVIRONMENTAL PROGRAM QUESTIONS

		Strongly Disagree	Partly Disagree	Don't Know	Partly Agree	Strongly Agree
1	My company has a published environmental policy or will have one in the next year.	1	2	3	4	5
2	My company has specific targets for improving its environmental performance.	1	2	3	4	5
3	My company publishes a meaningful annual environmental report.	1	2	3	4	5
4	My company has introduced, or is currently introducing, an environmental management system.	1	2	3	4	5
5	My company has introduced environmental considerations into its purchasing decisions.	1	2	3	4	5

6 My company provides employees with training regarding environmental issues.

   1   2   3   4   5

7 My company makes employees from all levels of the business responsible for company environmental performance.

   1   2   3   4   5

8 My company uses life cycle analysis as a means of assessing and minimizing the environmental impacts of its processes or products.

   1   2   3   4   5

9 My company's management understands the wider issues of sustainable development and is beginning to address them.

   1   2   3   4   5

10 My company plans to systematically reduce its dependence on fossil fuels.

   1   2   3   4   5

11 My company plans to systematically reduce its use (or manufacture) of toxic chemicals.

   1   2   3   4   5

12 My company plans to systematically reduce its consumption of unsustainable products. (For example: Forestry)

   1   2   3   4   5

13 My company applies the same high environmental standards to its activities at home and abroad.

   1   2   3   4   5

PART II - MANAGER SUPPORT FOR ENVIRONMENTAL INITIATIVE QUESTIONS

14. Have you ever tried to promote an environmental initiative within your company?

    Yes _____ No _____

	Unsupportive			*Supportive*	
	Highly	Partially	Didn't Care	Partially	Highly
15.If yes, how supportive was your manager of your environmental initiative(s)?	1	2	3	4	5

PART III - INSTRUCTIONS FOR BEHAVIORAL QUESTIONS

In the following section you are asked to answer questions about your manager's behavior. On each of the six pages that follow you are asked to do four things:

1. Using a box in the **"General"** column, **mark an "X" by one behavior** which you think best describes a **typical behavior** which you might expect to observe from your manager.

2. Then, using the **same list** of behaviors, mark an **"X" by one behavior** (using a box in the **"Environment"** column) which you think best describes a typical behavior you might expect from your manager with regards to his/her management of **environmental aspects of your job**.

3. **Assign a number from 1 (low) to 5 (high)** which represents the level of support which you would expect to experience from your manager.

4. **Describe one or more specific examples** of your manager's behaviors which either support or deter your participation in the area being addressed. Here you can choose to describe general or environmentally-specific examples, or both.

Example of how to complete this section:

Information Dissemination-Examples of Behaviors:	General	Environment
1. Neither actively aids nor hinders information flow to employees.	1.	1.
2. Uses information systems, such electronic bulletin boards, videos, computer systems, etc. to share information amongst employees.	2.	2.
3. If there is too much information, s/he sets priorities and establishes what is most important for employees to know.	3.	3.

If you feel that your manager generally does a good job of setting priorities when there is too much information, then you mark an **X in box #3** of the **"General" column**. If you think that your manager neither actively aids nor hinders the flow of environmentally-specific information to employees, then you would mark an **X in box #1** of the **"Environment" column**. Note that your manager may behave differently in general cases than in environmentally-specific cases. Please **choose only one behavior in each column.**

Choose a number between 1 (low) and 5 (high):
Place the number on the line provided.   **General   Environment**

_____   _____

3.5 = you think your manager is slightly above average in information sharing.
2.0 = you think your manager is below average in sharing environmental information.

Describe specific behaviors: "My manager uses regular group meetings to brief us on new initiatives and points out what is most important for us to know." **Or** "My manager seems indifferent with regards to sharing environmental information, acting like that information is irrelevant to our jobs."

A. Innovation: Mark an X in the **General** column by the **one behavior** which most accurately describes a typical experience you would expect with regards to your manager's encouragement (or discouragement) of innovation. Then mark an X in the **Environment** column by the **one behavior** which best describes your manager's encouragement (or discouragement) of innovations in the environmental area.

Examples of Behaviors	General	Environment
1. Usually objects to changes and new ideas and finds excuses why they can not be implemented.	1.	1.
2. Seldom experiments with new ideas or methods of doing things.	2.	2.
3. I would never approach my manager with a suggested change because I know s/he would be angry with me for interfering in his/her area of responsibility.	3.	3.
4. Neither encourages nor discourages new ideas from employees.	4.	4.
5. Gives feedback to employees on their ideas and suggestions, even if they are not adopted.	5.	5.
6. Would accompany an employee to discuss and promote the employee's idea to another manager.	6.	6.
7. Experiments with new ideas in order to examine whether they are profitable/feasible to adopt on a large scale.	7.	7.
8. When someone makes a mistake, we usually discuss, as a group, how to avoid the problem in the future.	8.	8.
9. Sends employees to other locations in the company and elsewhere to learn about innovative processes and other ways of doing business.	9.	9.
10. Encourages partnerships with other departments in order to implement new ideas.	10.	10.

	General	Environment
Choose a number between 1 (unsupportive) to 5 (highly supportive) which best describes the level of support you would expect from your manager with regards to innovation. Assign a separate number in each column.	_____	_____

Describe specific example(s) of how your manager has either supported or failed to support your ideas or your participation in innovations. (Answer in general or for environmentally-specific ideas or innovations.)

_____

_____

_____

B. Competence Building: Mark an X in the **General** column by the **one behavior** which most accurately describes a typical experience you would expect with regards to your manager's encouragement (or discouragement) of training, education and learning. Then mark an X in the **Environment** column by the **one behavior** which best describes your manager's encouragement (or discouragement) of environmental training, education and learning.

Examples of Behaviors	General	Environment
1. Refuses to commit resources and employee time for training and education activities.	1.	1.
2. Delays giving employees the training/education they need when they change their job functions.	2.	2.
3. Neither encourages nor discourages employee participation in training and education.	3.	3.
4. Usually encourages participation in any appropriate learning situation in which an employee would like to engage.	4.	4.
5. If there is something new I need to know, my manager will make sure I have training/ education on it quickly.	5.	5.
6. Realigns employee responsibilities to allow employee time for training, site visits, or exploring new techniques for doing his/her job.	6.	6.
7. Spends time discussing and implementing a learning plan with each employee.	7.	7.

	General	Environment
Choose a number from 1 (unsupportive) to 5 (highly supportive) which best describes the level of support you would expect from your manager with regards to competence building. Assign a separate number in each column.	_____	_____

Describe specific example(s) of how your manager has either supported or deterred your participation in training, education and/or learning. (Answer in general, or for environmentally-specific training, education and/or learning.)

_____

_____

_____

_____

_____

C. Communication: Mark an X in the **General** column by the **one behavior** which most accurately describes a typical experience you would expect with regards to your manager's encouragement (or discouragement) of open communication. Then mark an X in the **Environment** column by the **one behavior** which best describes your manager's encouragement (or discouragement) of communication concerning environmental issues.

Examples of Behaviors	General	Environment
1. Reinforces organizational hierarchies by insisting that employees be of the same level or the level immediately above in order to communicate. (i.e. Does not want employees to talk to other managers.)	1.	1.
2. Avoids difficult discussions and ignores problems as they are developing.	2.	2.
3. Listens to employees, then forgets/ignores what they have said.	3.	3.
4. Neither encourages nor discourages employee communication.	4.	4.
5. Encourages employees to express concerns about company decisions and policies so that the concerns can be openly discussed.	5.	5.
6. Creates an open environment in which to discuss decisions which affect the business. For example, welcomes employee discussions about possible changes, improvements or problems that need solving.	6.	6.
7. Answers questions honestly even if the answer is not what the employee wants to hear.	7.	7.
8. Listens to and values input from employees and managers from all parts of the company.	8.	8.
9. Listens openly and attentively to suggested improvements in how s/he does his/her job and often adopts the suggestions.	9.	9.

	General	Environment
Choose a number from 1 (unsupportive) to 5 (highly supportive) which describes the level of support you would expect from your manager with regards to communication. Assign a separate number in each column.	_____	_____

Describe specific example(s) of how your manager has either supported you or deterred you from communicating openly in general, or with regards to environmental concerns/ideas:

_____

_____

D. Information Dissemination: Mark an X in the **General** column by the **one behavior** which most accurately describes a typical experience you would expect with regards to your manager's openness and completeness in sharing company information. Then mark an X in the **Environment** column by the **one behavior** which best describes your manager's openness and completeness in sharing environmental information.

Examples of Behaviors	General	Environment
1. Our group is often the last to know about changes in the company because our manager does not tell us things.	1.	1.
2. Tries to hide unpopular decisions and information from employees.	2.	2.
3. Gives incomplete or inaccurate information to employees.	3.	3.
4. Keeps information about problems in our area private from the rest of the company and tries to solve them without help.	4.	4.
5. Neither actively aids nor hinders information flow to employees.	5.	5.
6. Uses information systems, such electronic bulletin boards, videos, computer systems, etc. to share information amongst employees.	6.	6.
7. If there is too much information, s/he sets priorities and establishes what is most important for employees to know.	7.	7.
8. Encourages employee trust by openly announcing information without delay about troubling situations, like lay-offs or restructurings.	8.	8.
9. Clearly explains the reason for the organization's goals or policies and forewarns employees about expected changes whenever possible.	9.	9.

	General	Environment
Choose a number from 1 (highly open/complete) to 5 (highly closed/incomplete) which best describes the level of openness and completeness you would expect from your manager with regards to information sharing. Assign a separate number in each column.	_____	_____

Describe specific example(s) of how your manager has either openly communicated information in a complete and timely manner or failed to do so. (Answer in general and/or for environment-specific information.)

_____

_____

E. Rewards and Recognition: Mark an X in the **General** column by the **one behavior** which most accurately describes a typical experience you would expect with regards to your manager's use of rewards and recognition. Then, mark an X in the **Environment** column by the **one behavior** which best describes your manager's use of rewards and recognition to encourage achievement of environmental goals.

Examples of Behaviors	General	Environment
1. I have seen my manager publicly reprimand another employee (or my manager has criticized me in front of others).	1.	1.
2. If I do a good job, I am not certain that my manager will notice. But, if I make a mistake, I am sure s/he will notice and probably criticize me for the mistake.	2.	2.
3. Seldom, if ever, rewards or recognizes an employee for work well done.	3.	3.
4. Neither recognizes nor discourages employee contributions.	4.	4.
5. If the company does well, my manager will reward all of his/her employees.	5.	5.
6. Looks for opportunities to praise positive employee performance, both privately and in front of others.	6.	6.
7. Rewards a good idea by implementing it and giving the responsible employee(s) credit.	7.	7.
8. Uses company award systems to recognize particularly good performance of employees.	8.	8.
9. Uses bonus pay or other monetary awards to reward employees who have achieved or surpassed their goals.	9.	9.

	General	Environment
Choose a number from 1 (seldom uses) to 5 (uses often and well) which best describes the level of use of rewards and recognition you would expect from your manager. Assign a separate number in each column.	_____	_____

Describe specific example(s) of how your manager has either used or failed to use rewards and recognition to demonstrate appreciation for your contributions to the company. (Answer in general and/or for environmentally-specific rewards and recognition.)

_____

_____

_____

_____

F. Management of Goals and Responsibilities: Mark an X in the **General** column by the **one behavior** which most accurately describes a typical experience you would expect with regards to your manager's monitoring and setting of goals and responsibilities. Then, mark an X in the **Environment** column by the **one behavior** which best describes your manager's setting of environmental goals and assigning of environmentally-specific responsibilities.

Examples of Behaviors	General	Environment
1. Tries to manage every detail of an employee's work so that the employee has little freedom to do his/her job independently.	1.	1.
2. Seldom talks to employees about goals and responsibilities, except when required by company policies.	2.	2.
3. Is often vague about what s/he wants from an employee.	3.	3.
4. Neither encourages or discourages employees from taking responsibility.	4.	4.
5. Keeps responsibility for all decisions.	5.	5.
6. Delegates specific tasks to employees and tells them precisely how the tasks should be performed.	6.	6.
7. Talks regularly with employees to assess progress toward explicit employee goals.	7.	7.
8. Tells an employee right away when there is something wrong with his/her work.	8.	8.
9. Uses both quantitative (numbers) and qualitative (quality) measures to assure individual is making progress toward or contributing to company goals.	9.	9.
10. Involves employees in changes by instilling ownership of problems and responsibilities for solutions in every employee.	10.	10.

	General	Environment
Choose a number from 1 (never sets goals /does not assign clear responsibilities) to 5 (strong use of goals /establishes clear responsibilities) which best describes the level of behavior you would expect from your manager. Assign a separate number in each column.	_____	_____

Describe specific example(s) of how your manager has either set goals /shared responsibilities with you for achieving company objectives or failed to do so, in general or in the environmental area.

_____

_____

Thank you for your participation.

Index

Printed and bound by CPI Group (UK) Ltd, Croydon, CR0 4YY

27/10/2024

01779847-0001